BIRKHÄUSER

Oberwolfach Seminars
Volume 40

Helga Baum
Andreas Juhl

Conformal Differential Geometry

Q-Curvature and Conformal Holonomy

Birkhäuser
Basel · Boston · Berlin

Authors:

Helga Baum
Humboldt-Universität
Institut für Mathematik
10099 Berlin
Germany
e-mail: baum@math.hu-berlin.de

Andreas Juhl
Humboldt-Universität
Institut für Mathematik
10099 Berlin
Germany
e-mail: ajuhl@math.hu-berlin.de

and

Universitet Uppsala
Matematiska Institutionen
Box-480
75106 Uppsala
Sweden
e-mail: andreasj@math.uu.se

2000 Mathematics Subject Classification 53A30, 53B15, 53B20, 53B25, 53B30, 53B50, 53C05, 53C10, 53C15, 53C21, 53C25, 53C27, 53C28, 53C29, 53C50, 53C80, 58 J50, 32V05

Library of Congress Control Number: 2009942367

Bibliographic information published by Die Deutsche Bibliothek
Die Deutsche Bibliothek lists this publication in the Deutsche Nationalbibliografie;
detailed bibliographic data is available in the Internet at <http://dnb.ddb.de>.

ISBN 978-3-7643-9908-5 Birkhäuser Verlag, Basel – Boston – Berlin

Basel · Boston · Berlin
P.O. Box 133, CH-4010 Basel, Switzerland
Part of Springer Science+Business Media
Printed on acid-free paper produced from chlorine-free pulp. TCF ∞
Printed in Germany

ISBN 978-3-7643-9908-5 e-ISBN 978-3-7643-9909-2

9 8 7 6 5 4 3 2 1 www.birkhauser.ch

Contents

Preface

This volume is based on lectures that were presented during the Oberwolfach-Seminar on Recent Developments in Conformal Differential Geometry at the *Mathematisches Forschungsinstitut Oberwolfach* MFO in November 2007. It consists of two main parts: one on Q-curvature and one on conformal holonomy. Both parts are independent of each other and reflect recent developments of the corresponding topics.

In the early 1990s, Branson [Br93] introduced the concept of Q-curvature in connection with an investigation of the behaviour of functional determinants of natural conformally covariant operators under conformal changes of the Riemannian metric. The underlying idea was that Q-curvature should define a universal part of the conformal anomalies of all such determinants. The full realization of that idea remains a challenge. Since then the notion of Q-curvature has become central in conformal geometry and geometric analysis: see [C04], [C05] and the references therein. Moreover, it has a natural place in the study of the interplay of spectral theory on asymptotically hyperbolic manifolds and the geometry of their conformal infinities (in the sense of Penrose). In recent years, the systematic exploration of these relations has been much stimulated by the advent of the AdS/CFT-duality [M98], [Wi98]. This perspective naturally fits also with the program initiated by Fefferman and Graham in [FG85]. The seminal work [GZ03] was one of the fruits of these interactions. Among other things, it established a link between the total Q-curvature of conformal infinity and the total holographic anomaly of the renormalized volume. In the recent monograph [J09b], these questions were treated from a new perspective, which systematically regards previous results as part of a theory of conformally covariant differential operators that are associated to hypersurfaces in conformal manifolds. One of the new findings of that theory is that of recursive structures among Q-curvatures and the associated conformally covariant powers of the Laplacian. In the present lectures on Q-curvature we motivate its study, describe its current role, and introduce the ideas of [J09b].

The concept now called Cartan connections was introduced and successfully applied by E. Cartan during the first decades of the 20th century. Cartan studied geometric structures associated with solution spaces of classical ordinary differential equations using moving frames. Thereby, Cartan found a uniform concept that, by studying invariants of group actions, as well as B. Riemann's concept of curved

Riemannian manifolds, generalizes F. Klein's approach to geometry to a broad class of geometries, such as conformal, projective, Einstein-Weyl, 3-dimensional CR and so on. The renewed interest in conformal, projective and quaternionic structures as well as relations to complex analysis via CR structures brought Cartan connections back to general interest during the last decade. In particular, the systematic development of parabolic differential geometry in modern language of principle fibre bundles (cf. [CS09]) turned out to be extremely useful in studying geometric problems and structures in a unified and systematic way. We mention here in particular the classification of various conformal and CR invariants, the classification of conformally invariant operators, the classification of higher symmetries of differential operators, the classification of solutions of twistor equations, or the progress in understanding T. Branson's Q-curvature (cf. for example [FG85], [CSS01], [CD01], [E05], [L07], [J09b]).

The common feature of many different kinds of geometries is the existence of a distinguished (normal) Cartan connection which characterizes the geometry in question, similar to the Levi-Civita connection in Riemannian geometry. An example is conformal geometry, which we will focus on in the second chapter of this volume. Many symmetries and supersymmetries in conformal geometry (conformal vector fields, conformal Killing forms, conformal Killing tensors, conformal Killing spinors) have an interpretation as invariant objects for the holonomy group of the normal conformal Cartan connection and can be classified in this way. In the second part of this volume, we want to introduce the reader to this application of Cartan geometry. We will explain the normal conformal Cartan connection and the holonomy group of a conformal manifold. In particular we intend to show how geometries admitting conformal symmetries and supersymmetries can be classified using methods of conformal Cartan geometry.

The original lectures were designed as introductions to the respective subjects with particular emphasis on the underlying conceptual ideas rather than on technical details. The present notes are an expanded version of the lectures containing more details and describing more background. In particular, we give proofs for a selection of central facts. Since the presentation in Chapter 1 does not overlap much with other texts and also contains some new results, these lectures might also have some interest for experts. In our presentation of conformal Cartan geometry in Chapter 2, we try to avoid the general machinery of parabolic geometry. Our main aim is to explain the link between the conformally invariant concepts of conformal Cartan geometry and concrete conformally invariant geometric structures on (pseudo-)Riemannian manifolds as explicit as possible.

The list of references is far from being exhaustive. We have restricted citations to only those sources which are of immediate concern for these lectures. We apologize for all serious omissions.

Chapter 1 starts with a brief discussion of the flat model of conformal geometry in Section 1.1. After these preparations, Section 1.2 introduces and discusses Q-curvature of order 4 in arbitrary dimension. Among other things, we prove the conformal covariance of the Paneitz operator, describe the role of Q_4 in the conformal variation of functional determinants (Polyakov's formula) and discuss its connection with the Gauß-Bonnet formula. In Section 1.3, we define Q-curvature in full generality and derive its fundamental transformation law. Then we use representation theory to derive explicit formulas for GJMS-operators and Q-curvatures on the round sphere. We identify GJMS-operators with the residues of an intertwining operator for spherical principal series representations. This serves as a preparation for Section 1.4, which explores the connection to scattering theory on conformally compact Einstein spaces. Again, we begin with a detailed description of that perspective for the round sphere. It is followed by showing that the critical Q-curvature and the GJMS-operators for general metrics can be seen in the scattering operator. This material is crucial for the last two sections which are devoted to a discussion of some results of [J09b]. The main ideas here are the construction of the so-called residue families and the associated Q-curvature polynomials. The residue families are certain families of differential operators which may be regarded as local analogs of the scattering operator. In particular, the critical family contains the critical Q-curvature and the critical GJMS-operator in much the same way as the scattering operator contains these data, and the parameter of the residue families corresponds to the spectral parameter of the scattering operator. Section 1.6 describes recursive structures for Q-curvatures and GJMS-operators. These are consequences of their relation to the residue families. In particular, we display well-structured explicit formulas for the conformally covariant cube P_6 of the Laplacian and the Q-curvature Q_6. We hope that the latter results are amenable to future applications in geometric analysis.

Chapter 2 is organized as follows. In Section 2.1 we introduce the notion of a Cartan connection and of its holonomy group. In particular, we relate this holonomy group to the holonomy group of covariant derivatives on associated tractor bundles. Section 2.2 contains an outline of the construction of the normal conformal Cartan connection in its invariant form as well as its description in terms of the metrics in the conformal class. This leads to the 'metric' formula for the normal covariant derivative on tractor bundles, which is usually used in conformal geometry. In particular, in conformal geometry the normal covariant derivative on the standard tractor bundle of a conformal manifold plays the same role as the Levi-Civita connection plays in (pseudo)-Riemannian geometry. Due to this, the holonomy group of a conformal manifold is defined as the holonomy group of this distinguished conformally invariant covariant derivative. Section 2.3 is devoted to conformal manifolds with reducible conformal holonomy representation. It turns out that this case is closely related to the existence of Einstein metrics or products of Einstein metrics in the conformal class. In particular, we describe the conformal analog of de Rham's Splitting Theorem for conformal manifolds. In Section 2.4 we give a review of recent classification results for irreducibly acting conformal holonomy groups in Riemannian and in Lorentzian signatures. In Section 2.5 and Section 2.6 we intend to show how twistor equations on differential forms and on spinors can be solved using conformal holonomy groups. We focus mainly on the case of spinors, where we also give an overview of classification results for the geometric structures admitting conformal Killing spinors in the Riemannian as well as in the Lorentzian case. In the latter case, methods of conformal Cartan geometry proved to be extremely useful. Section 2.7 deals with Lorentzian manifolds with conformal holonomy group in $\mathrm{SU}(1,m)$, the "conformal analog of Calabi-Yau manifolds". For this aim, we explain the construction of Fefferman spaces, which are S^1-bundles over strictly pseudoconvex CR manifolds. Using results about conformal Killing spinors on Fefferman spaces, we explain why the conformal holonomy group of Fefferman spaces is contained in $\mathrm{SU}(1,m)$. We finish part two of this volume with an outlook on recent results on conformal manifolds with holonomy groups in $\mathrm{SU}(p,q)$, $\mathrm{Sp}(p,q)$, $G_{2(2)}$ and $\mathrm{Spin}(4,3)$.

Acknowledgments. We are grateful to MFO and its staff for providing the infrastructure which enabled realization of the seminar on conformal geometry. We thank MFO's director Gert-Martin Greuel for urging us to publish these lecture notes. During the actual preparation of the manuscript we enjoyed an effective cooperation with the Birkhäuser editor Thomas Hempfling and his team. Finally, we are grateful to Carsten Falk for critical reading of parts of the manuscript. The second author was supported by a DFG grant through SFB 647 "Space-Time-Matter".

Chapter 1

Q-curvature

The concept of Q-curvature was introduced by Thomas Branson in [Br93], [Br95] and [BO91b]. Since then Q-curvature has played an increasing and central role in large parts of conformal differential geometry and related geometric analysis. For recent reviews see [Br05], [BG08], [CEOY08] and [M07].

1.1 The flat model of conformal geometry

We start by describing the flat model of conformal geometry. Let $n \geq 3$. The Euclidean space $\mathbb{R}^{n+1}$ with the scalar product

$$(x, y) = -x_0 y_0 + \sum_{i=1}^{n} x_i y_i, \quad x = (x_0, x'), \ y = (y_0, y')$$

will be denoted by $\mathbb{R}^{1,n}$. Let $b(x) = -x_0^2 + \sum_{i=1}^{n} x_i^2$, and let $\mathrm{O}(1, n)$ be the orthogonal group of the quadratic form b, i.e.,

$$\mathrm{O}(1, n) = \left\{ T \in \mathrm{GL}(n+1, \mathbb{R}) \mid b(Tx) = b(x), \ x \in \mathbb{R}^{n+1} \right\}.$$

Elements in $\mathrm{O}(1, n)$ will be written in block form with respect to natural bases of $\mathbb{R}^{n+1}$. The standard basis $\{e_i\}_0^n$ corresponds to block matrices of the form

$$\begin{pmatrix} 1 \times 1 & 1 \times n \\ n \times 1 & n \times n \end{pmatrix}.$$

In addition, it will be convenient to use the basis

$$\{f_0, f_1, \ldots, f_{n-1}, f_n\} = \{f_0, e_1, \ldots, e_{n-1}, f_n\} \tag{1.1.1}$$

with

$$f_0 = \frac{e_n - e_0}{\sqrt{2}} \quad \text{and} \quad f_n = \frac{e_n + e_0}{\sqrt{2}}.$$

Note that $(f_0, f_0) = (f_n, f_n) = 0$. Then

$$(b(f_i, f_j)) = \begin{pmatrix} 0 & & 1 \\ & I_{n-1} & \\ 1 & & 0 \end{pmatrix},$$

and matrices in $\mathrm{O}(1, n)$ have the form

$$\begin{pmatrix} 1\times 1 & & 1\times 1 \\ & n-1\times n-1 & \\ 1\times 1 & & 1\times 1 \end{pmatrix}.$$

Next, we define some subgroups of $\mathrm{O}(1, n)$ which will play a distinguished role. With respect to the basis (1.1.1) let

$$A = \left\{ \begin{pmatrix} a^{-1} & & 0 \\ & I & \\ 0 & & a \end{pmatrix}, \ a \in \mathbb{R}, \ a \neq 0 \right\}$$

and

$$M = \left\{ \begin{pmatrix} 1 & & \\ & T & \\ & & 1 \end{pmatrix}, \ T \in \mathrm{O}(n-1) \right\}.$$

Moreover, let

$$N^- = \left\{ \begin{pmatrix} 1 & v & -\frac{1}{2}b(v) \\ & I & -v \\ 0 & & 1 \end{pmatrix}, \ v \in \mathbb{R}^{n-1} \right\}$$

and

$$N^+ = \left\{ \begin{pmatrix} 1 & & 0 \\ v & I & 0 \\ -\frac{1}{2}b(v) & -v & 1 \end{pmatrix}, \ v \in \mathbb{R}^{n-1} \right\}.$$

The real Lie algebras of these groups will be denoted by $\mathfrak{a}$, $\mathfrak{m}$ and $\mathfrak{n}^{\pm}$.

The basic properties of these groups are the following. First of all, A and $N^{\pm}$ are abelian and A and M commute. Next, the group MA normalizes N^+ and N^-. Hence the groups

$$P^{\pm} = MAN^{\pm}$$

are well-defined. Let $\mathfrak{p}^{\pm} = \mathfrak{m} \oplus \mathfrak{a} \oplus \mathfrak{n}^{\pm}$ be the Lie algebra of $P^{\pm}$. $\mathfrak{m} \oplus \mathfrak{a}$ and $\mathfrak{n}^{\pm}$ give rise to the decomposition

$$\mathfrak{g} = \mathfrak{n}^- \oplus \mathfrak{m} \oplus \mathfrak{a} \oplus \mathfrak{n}^+ \tag{1.1.2}$$

of the Lie algebra $\mathfrak{g}$ of $G = \mathrm{O}(1, n)$. (1.1.2) is the root decomposition of $\mathfrak{g}$ with respect to the adjoint action of $\mathfrak{a}$, i.e.,

$$\mathrm{ad}(X_0), \ X_0 = \begin{pmatrix} -1 & & \\ & 0 & \\ & & 1 \end{pmatrix} \in \mathfrak{a}$$

acts with the eigenvalues ± 1 on $\mathfrak{n}^{\pm}$. In particular,

$$[\mathfrak{n}^+, \mathfrak{n}^-] \subset \mathfrak{m} \oplus \mathfrak{a}.$$

Note also that the adjoint action of $M \simeq \mathrm{O}(n-1)$ on $\mathfrak{n}^{\pm} \simeq \mathbb{R}^{n-1}$ coincides with the standard action.

In Chapter 2, the alternative notation

$$B = P^-, \quad N^- = B_1, \quad N^+ = B_{-1} \quad B_0 = MA$$

and

$$\mathfrak{b}_1 = \mathfrak{n}^-, \quad \mathfrak{b}_{-1} = \mathfrak{n}^+, \quad \mathfrak{b}_0 = \mathfrak{m} \oplus \mathfrak{a}$$

will be used. Then (1.1.2) reads

$$\mathfrak{g} = \mathfrak{b}_1 \oplus \mathfrak{b}_0 \oplus \mathfrak{b}_{-1},$$

and we have $[\mathfrak{b}_i, \mathfrak{b}_j] \subset \mathfrak{b}_{i+j}$, i.e., $\mathfrak{g}$ is a 1-graded Lie algebra.

Now the action of $\mathrm{O}(1,n)$ preserves the light-cone

$$C = \{x = (x_0, x') \in \mathbb{R}^{1,n} \,|\, b(x) = 0\}.$$

Since the linear action of $\mathrm{O}(1,n)$ on $\mathbb{R}^{1,n}$ commutes with the action $x \mapsto \lambda x$, $\lambda \in \mathbb{R}$, it induces an action on the space of lines in $\mathbb{R}^{n+1}$, i.e., on the real projective space $\mathbb{P}^n$. Let $Q \subset \mathbb{P}^n$ be the space of lines in the cone C, and let $\pi : C \to Q$ be the natural projection.

Through the identification

$$S^{n-1} \ni x \mapsto \pi((1,x)) \in Q \tag{1.1.3}$$

the group $\mathrm{O}(1,n)$ acts by

$$x \mapsto \pi((1,x)) \mapsto \pi(g(1,x)) = \pi\left(\left(1, \frac{g(1,x)'}{g(1,x)_0}\right)\right) \mapsto \frac{g(1,x)'}{g(1,x)_0} \tag{1.1.4}$$

on the sphere S^{n-1}. In order to determine an explicit formula for the action, let $g = \begin{pmatrix} d & c \\ b^t & A \end{pmatrix}$ (with arrows b, c of length n and a scalar d); here we use the standard basis. Then $g(1,x) = (d + (c,x), b + xA^t)$, and g induces the map

$$x \mapsto \frac{xA^t + b}{(c,x) + d} \tag{1.1.5}$$

on S^{n-1}.

The space $\mathbb{R}^{1,n}$ is a Lorentzian manifold with the metric

$$\tilde{g} = -dx_0^2 + \sum_{i=1}^{n} dx_i^2. \tag{1.1.6}$$

The restriction of $\tilde{g}$ to the cone C degenerates. In fact, $\tilde{g}(X,Y) = 0$ for $X = \sum_{i=0}^n x_i \partial/\partial x_i \in T_x(C)$ and all $Y \in T_x(C)$. This follows from $db(Y) = 2\tilde{g}(X,Y)$. But for any $x \in C$,

$$g_x(Y,Z) = g_x(\pi^{-1}(Y), \pi^{-1}(Z)),\ Y,Z \in T_{\pi(x)}(Q)$$

defines an inner product on $T_{\pi(x)}(Q)$. Moreover, we have $g_{\lambda x} = \lambda^2 g_x$. In other words, any section $\eta : Q \to C$ of $\pi : C \to Q$ induces a metric $\eta^*(\tilde{g})$ on Q so that two such sections induce metrics which differ by a non-vanishing smooth function, i.e., are conformally equivalent (see Definition 1.2.1). This defines a conformal class on Q. Through the identification (1.1.3) of Q with S^{n-1}, this class is the conformal class of the round metric g_c.

Now $\mathrm{O}(1,n)$ operates by conformal diffeomorphisms with respect to g_c, i.e., for any $g \in \mathrm{O}(1,n)$, there is a non-vanishing function $\Phi_g \in C^\infty(S^{n-1})$ so that

$$g_*(g_c) = \Phi_g^2 g_c.$$

More precisely, (1.1.4) implies that

$$g_*(g_c) = \frac{1}{(d-(b,x))^2} g_c, \quad g = \begin{pmatrix} d & c \\ b^t & A \end{pmatrix}. \tag{1.1.7}$$

$\mathrm{O}(1,n)$ acts transitively on Q with isotropy group $G_{\pi(f_0)} = P^-$. Thus $Q \simeq \mathrm{O}(1,n)/P^-$. In terms of the sphere S^{n-1}, the parabolic subgroup P^- fixes the point $(0,0,\dots,-1)$.

Note that the action (1.1.5) of $\mathrm{O}(1,n)$ on S^{n-1} extends the *isometric* action on the unit ball $\mathbb{B}^n = \{x \in \mathbb{R}^n,\ |x| < 1\}$ with the metric

$$\frac{4}{(1-|x|^2)^2} \sum_{i=1}^n dx_i^2. \tag{1.1.8}$$

In order to see that, we use the fact that

$$C(-1) \ni (x_0, x') \mapsto \frac{x'}{1+x_0} \in \mathbb{B}^n$$

is an isometry of

$$C(-1) = \{x = (x_0, x') \in \mathbb{R}^{1,n} \mid b(x) = -1\}$$

and $\mathbb{B}^n$. Its inverse is given by

$$\mathbb{B}^n \ni x \mapsto \left(\frac{1+|x|^2}{1-|x|^2}, \frac{2x}{1-|x|^2} \right) \in C(-1).$$

These formulas show that $\begin{pmatrix} d & c \\ b^t & A \end{pmatrix}$ operates on $\mathbb{B}^n$ by

$$x \mapsto \frac{2xA^t + b(1+|x|^2)}{1-|x|^2 + 2(c,x) + d(1+|x|^2)}. \tag{1.1.9}$$

For $x \in S^{n-1} = \partial \mathbb{B}^n$, the latter formula reproduces (1.1.5).

Similarly, the quadratic form

$$b(x) = -\sum_{i=0}^{p} x_i^2 + \sum_{i=p+1}^{p+q+1} x_i^2 \tag{1.1.10}$$

on $\mathbb{R}^{p+1,q+1}$ defines a quadric $Q^{p,q}$ in the projective space $\mathbb{P}^n$, $n = p+q+1$. The group $\mathrm{O}(p+1, q+1)$ acts on $Q^{p,q}$ by linear transformations. $Q^{p,q}$ is a homogeneous space for $\mathrm{O}(p+1, q+1)$. The isotropy group P^- at $\pi(f_0)$ has the decomposition $P^- = MAN^-$, where the groups M, A and N^- are given by

$$A = \left\{ \begin{pmatrix} a^{-1} & & 0 \\ & I_{p+q} & \\ 0 & & a \end{pmatrix}, \ a \in \mathbb{R}, \, a \neq 0 \right\}, \quad M = \left\{ \begin{pmatrix} 1 & & \\ & T & \\ & & 1 \end{pmatrix}, \ T \in \mathrm{O}(p,q) \right\}$$

and

$$N^- = \left\{ \begin{pmatrix} 1 & v & -\frac{1}{2}b(v) \\ & I_{p+q} & -J_{p,q}v^t \\ 0 & & 1 \end{pmatrix}, \ v \in \mathbb{R}^{p,q} \right\}, \quad J_{p,q} = \begin{pmatrix} -I_p & \\ & I_q \end{pmatrix}.$$

The corresponding Lie algebras $\mathfrak{a}$, $\mathfrak{m}$, $\mathfrak{n}^{\pm}$ yield a generalization of the decomposition (1.1.2) for the Lie algebra of $\mathrm{O}(p+1, q+1)$. In particular, $\mathfrak{g}$ is 1-graded.

The signature $(p+1, q+1)$ metric

$$\tilde{g} = -\sum_{i=0}^{p} dx_i^2 + \sum_{i=p+1}^{p+q+1} dx_i^2 \tag{1.1.11}$$

on $\mathbb{R}^{p+1,q+1}$ induces a conformal class of metrics of signature (p,q) on $Q^{p,q}$. Sometimes this space is called the Möbius sphere. In the special case $p = 1$, we obtain an n-dimensional Lorentzian manifold $Q^{1,n-1}$ with a double cover given by the product space $S^1 \times S^{n-1}$ with the metric $-g_{S^1} + g_{S^{n-1}}$.

For more details on the flat model we refer to [Fe05], [Sh97] and [F07].

1.2 Q-curvature of order 4

In this lecture we define the Q-curvature Q_4 and discuss its basic properties.

The quantity Q_4 is a scalar curvature quantity of order 4 of a smooth Riemannian manifold of dimension $n \geq 3$, i.e., its definition involves four derivatives of the metric. Among other curvature quantities of the same order, Q_4 is distinguished by its behaviour with respect to conformal changes of the metric.

First, we make precise the notions of conformal change and conformal class.

Definition 1.2.1. The Riemannian metrics g and g' on M are called conformally equivalent iff $g = \lambda g'$ for some positive $\lambda \in C^\infty(M)$. The set of all metrics of the form λg with $0 < \lambda \in C^\infty(M)$ is called the conformal class $[g]$ of g.

It will be convenient to write conformal factors in the form

$$\lambda = e^{2\varphi}, \ \varphi \in C^\infty(M).$$

For given g, the conformally equivalent metrics will often be denoted by $\hat{g}$. In situations where this notation is used, the conformal factors should be clear from the context.

Next, we recall the transformation rules of the basic curvature tensors of Riemannian geometry with respect to conformal changes of the metric. For any g, let

$$\nabla^g : \Gamma(TM) \to \Gamma(T^*M \otimes TM) \tag{1.2.1}$$

denote the corresponding Levi-Civita connection. It defines a covariant derivative

$$\nabla^g_X : \mathfrak{X}(M) \to \mathfrak{X}(M) \tag{1.2.2}$$

on the vector space $\Gamma(TM) = \mathfrak{X}(M)$ of smooth vector fields on M so that

$$Xg(Y,Z) = g(\nabla^g_X(Y), Z) + g(Y, \nabla^g_X(Z)) \text{ and } \nabla^g_X(Y) - \nabla^g_Y(X) = [X,Y] \tag{1.2.3}$$

for all vector fields $X, Y, Z \in \mathfrak{X}(M)$. Let

$$R^g(X,Y) = \nabla^g_X \nabla^g_Y - \nabla^g_Y \nabla^g_X - \nabla^g_{[X,Y]} \tag{1.2.4}$$

be the Riemann curvature tensor. In terms of a local orthonormal basis $\{e_i\}$ we define the *Ricci tensor*

$$\mathrm{Ric}^g(X,Y) = \sum_i g(R^g(X,e_i)e_i, Y) \tag{1.2.5}$$

and the *scalar curvature*

$$\mathrm{scal}^g = \sum_i \mathrm{Ric}^g(e_i, e_i). \tag{1.2.6}$$

We shall also write $\mathrm{Ric}(g)$ and $\mathrm{scal}(g)$.

Let the gradient $\mathrm{grad}^g(f) \in \mathfrak{X}(M)$ of $f \in C^\infty(M)$ be defined by

$$g(\mathrm{grad}^g(f), X) = \langle df, X\rangle \quad \text{for all } X \in \mathfrak{X}(M),$$

and let

$$\mathrm{Hess}^g(f)(X,Y) = g(\nabla^g_X(\mathrm{grad}(f), Y), \ X, Y \in \mathfrak{X}(M) \tag{1.2.7}$$

be the (covariant) *Hessian form* of f. The trace of the symmetric bilinear form $\mathrm{Hess}^g(f)$ yields the Laplace-Beltrami operator for the metric g:

$$\Delta_g(f) = \mathrm{tr}_g \, \mathrm{Hess}^g(f). \tag{1.2.8}$$

Now the conformal transformation law

$$\mathrm{Ric}(e^{2\varphi}g) = \mathrm{Ric}(g) - (n-2)\,\mathrm{Hess}^g(\varphi) - \Delta_g(\varphi)g \\ - (n-2)|\,\mathrm{grad}^g(\varphi)|^2 g + (n-2)d\varphi \otimes d\varphi \quad (1.2.9)$$

for the Ricci tensor implies the conformal transformation law

$$e^{2\varphi}\mathrm{scal}(e^{2\varphi}g) = \mathrm{scal}(g) - (2n-2)\Delta_g(\varphi) - (n-1)(n-2)|\,\mathrm{grad}^g(\varphi)|^2 \quad (1.2.10)$$

for the scalar curvature.

For $n = 2$, these formulas read as

$$\mathrm{Ric}(e^{2\varphi}g) = \mathrm{Ric}(g) - \Delta_g(\varphi)g \quad \text{and} \quad e^{2\varphi}\mathrm{scal}(e^{2\varphi}g) = \mathrm{scal}(g) - 2\Delta_g(\varphi). \quad (1.2.11)$$

The second relation is equivalent to the transformation formula

$$e^{2\varphi}K(e^{2\varphi}g) = K(g) - \Delta_g(\varphi) \quad (1.2.12)$$

for the *Gauß-curvature*

$$K(g) = g(R(e_1, e_2)e_2, e_1)$$

of the surface M. The relation (1.2.12) is known as the *Gauß-curvature prescription equation.*

For $n \geq 3$, it is convenient to formulate (1.2.9) and (1.2.10) in terms of the *Schouten tensor*

$$\mathsf{P}^g = \frac{1}{n-2}\left(\mathrm{Ric}^g - \mathsf{J}^g g\right), \quad (1.2.13)$$

where

$$\mathsf{J}^g = \mathsf{J}(g) = \frac{\mathrm{scal}(g)}{2(n-1)}. \quad (1.2.14)$$

Then

$$\hat{\mathsf{P}} = \mathsf{P} - \mathrm{Hess}(\varphi) + d\varphi \otimes d\varphi - \frac{1}{2}|\,\mathrm{grad}(\varphi)|^2 g \quad (1.2.15)$$

and

$$e^{2\varphi}\hat{\mathsf{J}} = \mathsf{J} - \Delta(\varphi) - \left(\frac{n}{2}-1\right)|\,\mathrm{grad}(\varphi)|^2. \quad (1.2.16)$$

Here we abbreviate P and J for the metric $\hat{g} = e^{2\varphi}g$ by $\hat{\mathsf{P}}$ and $\hat{\mathsf{J}}$, respectively. This convention will be used throughout. Note that $\mathsf{J} = K$ in dimension 2.

The fundamental role of the Schouten tensor P in conformal differential geometry rests on the following decomposition formula for the curvature tensor (see Section 1G in [Be87]).

Proposition 1.2.1. *The curvature tensor of Riemannian manifolds of dimension $n \geq 3$ admits a decomposition of the form*

$$R^g = \mathsf{W}^g - \mathsf{P}^g \owedge g,$$

with a tensor W so that

$$\hat{\mathsf{W}} = e^{2\varphi}\mathsf{W}. \quad (1.2.17)$$

W is called the Weyl-tensor.

Here $\owedge$ denotes the *Kulkarni-Nomizu product* of symmetric bilinear forms. It is defined by

$$(b_1 \owedge b_2)(X,Y,Z,W) = b_1(X,Z)b_2(Y,W) - b_1(Y,Z)b_2(X,W) + b_1(Y,W)b_2(X,Z) - b_1(X,W)b_2(Y,Z).$$

In dimension $n \geq 4$, the vanishing of the Weyl-tensor W characterizes local conformal flatness. Thus, Proposition 1.2.1 implies that for locally conformally flat metrics in these dimensions the curvature tensor is governed by the Schouten tensor.

A Riemannian metric induces natural scalar products and norms on tensors. In particular, (1.2.17) implies

$$e^{4\varphi}\hat{g}(\hat{\mathsf{W}},\hat{\mathsf{W}}) = g(\mathsf{W},\mathsf{W}). \tag{1.2.18}$$

Using $\mathrm{vol}(e^{2\varphi}g) = e^{4\varphi}\,\mathrm{vol}(g)$, it follows that on closed four-manifolds the functional

$$\mathcal{W}(g) = \int_{M^4} g(\mathsf{W}^g,\mathsf{W}^g)\,\mathrm{vol}(g) = \int_{M^4} |\mathsf{W}^g|^2\,\mathrm{vol}(g) \tag{1.2.19}$$

is conformally invariant:

$$\mathcal{W}(e^{2\varphi}g) = \mathcal{W}(g).$$

g is a critical metric for the functional $\mathcal{W}$ iff g is Bach-flat, i.e., $\mathcal{B}(g) = 0$. Here

$$\mathcal{B}_{ij} = \nabla^l\nabla^k(\mathsf{W})_{kijl} + \frac{1}{2}\,\mathrm{Ric}^{kl}\,\mathsf{W}_{kijl} \tag{1.2.20}$$

is the *Bach tensor*.

After these preparations, we introduce Q_4 in general dimensions.

Definition 1.2.2 (Q-curvature of order four). On manifolds of dimension $n \geq 3$, we define Branson's Q-curvature of order four by

$$Q_{4,n} = \frac{n}{2}\mathsf{J}^2 - 2|\mathsf{P}|^2 - \Delta\mathsf{J}. \tag{1.2.21}$$

On the right-hand side of (1.2.21), all quantities are to be understood with respect to a fixed metric. In order to simplify the formula, the metric is suppressed. Due to the contribution $\Delta\mathsf{J}$ the definition of Q_4 involves four derivatives of the metric.

Of course, $Q_{4,n}$ can be written also in terms of more conventional quantities as the trace-free Ricci-tensor Ric_0 and the scalar curvature. Then we have the less illuminating formula

$$Q_{4,n} = c_n|\,\mathrm{Ric}_0\,|^2 + d_n\mathrm{scal}^2 - \frac{1}{2(\mathrm{n}-1)}\Delta\mathrm{scal}$$

with the constants

$$c_n = -\frac{2}{(n-2)^2} \quad \text{and} \quad d_n = \frac{(n-2)(n+2)}{8n(n-1)^2}.$$

Of particular interest will be Q_4 on manifolds of dimension $n = 4$. We set

$$Q_4 = Q_{4,4}$$

and call Q_4 the *critical* Q-curvature. Then

$$Q_4 = 2(\mathsf{J}^2 - |\mathsf{P}|^2) - \Delta \mathsf{J} \tag{1.2.22}$$

or, equivalently,

$$Q_4 = -\frac{1}{2}|\,\mathrm{Ric}_0\,|^2 + \frac{1}{24}\mathrm{scal}^2 - \frac{1}{6}\Delta \mathrm{scal}.$$

Later it will often be convenient to simply write Q_4 for all Q-curvatures $Q_{4,n}$. In such cases, the dimension n will be clear from the context.

The following result is the conformal transformation law of the critical Q-curvature Q_4. It was discovered in [BO91b].

Theorem 1.2.1 (The fundamental identity). *On four-manifolds* (M, g),

$$e^{4\varphi} Q_4(e^{2\varphi} g) = Q_4(g) + P_4(g)(\varphi), \tag{1.2.23}$$

where

$$P_4(g) = \Delta_g^2 + \delta_g(2\mathsf{J}g - 4\mathsf{P}^g)d. \tag{1.2.24}$$

In (1.2.24), the symmetric bilinear form P is considered as a linear operator on $\Omega^1(M)$, and the divergence $\delta_g : \Omega^1(M) \to C^\infty(M)$ is the adjoint of the exterior derivative with respect to the usual scalar products on differential forms induced by g.

In connection with Theorem 1.2.1 some comments are in order.

Since the definition of the curvature quantity Q_4 involves four derivatives of the metric, it is natural to expect that its transformation law under conformal changes $g \mapsto e^{2\varphi} g$ of the metric involves non-linearities in φ of order up to 4. However, Theorem 1.2.1 states that the transformation law of Q_4 is governed by a *linear* operator.

The situation should be compared with the following observation for scalar curvature. The transformation law (1.2.10) of the second-order scalar curvature displays non-linearities of order 2 if $n \geq 3$. However, in two dimensions, the corresponding transformation law in (1.2.11) is *linear* in φ. Thus, (1.2.23) should be regarded as an analog of the Gauß-curvature prescription equation (1.2.12). Therefore, it is sometimes called the Q_4-curvature prescription equation.

The transformation law of the critical Q-curvature Q_4 is governed by the *critical* Paneitz operator P_4. The fact that the operator P_4 appears in the conformal transformation law of Q_4 implies its conformal covariance. More precisely,

Theorem 1.2.2. *On four-manifolds (M,g),*

$$e^{4\varphi}P_4(e^{2\varphi}g) = P_4(g).$$

Proof. Theorem 1.2.1 implies

$$\begin{aligned} e^{4\varphi}P_4(e^{2\varphi}g)(\psi) &= e^{4\varphi}\left(e^{4\psi}Q_4(e^{2(\varphi+\psi)}g) - Q_4(e^{2\varphi}g)\right) \\ &= e^{4(\varphi+\psi)}Q_4(e^{2(\varphi+\psi)}g) - e^{4\varphi}Q_4(e^{2\varphi}g). \end{aligned}$$

Another application of Theorem 1.2.1 shows that the latter sum equals

$$Q_4(g) - P_4(g)(\varphi+\psi) - (Q_4(g) + P_4(g)(\varphi)) = P_4(g)(\psi).$$

This completes the proof. □

The same argument as in the proof of Theorem 1.2.2 derives the conformal covariance

$$e^{2\varphi}P_2(e^{2\varphi}g) = P_2(g) \tag{1.2.25}$$

of the Laplacian $P_2(g) = \Delta_g$ on a surface from the Gauß-curvature prescription equation (1.2.12). In order to further stress the analogy, we set $Q_2 = \mathsf{J}$ (in all dimensions).

It is an important fact that, for any metric g on a closed four-manifold M, the total Q-curvature

$$\int_{M^4} Q_4(g)\,\mathrm{vol}(g)$$

is conformally invariant. This is an easy consequence of Theorem 1.2.1. In fact, we calculate

$$\begin{aligned} \int_{M^4} Q_4(\hat{g})\,\mathrm{vol}(\hat{g}) &= \int_{M^4} e^{4\varphi}Q_4(\hat{g})\,\mathrm{vol}(g) \\ &= \int_{M^4} [Q_4(g) + P_4(g)(\varphi)]\,\mathrm{vol}(g) \\ &= \int_{M^4} Q_4(g)\,\mathrm{vol}(g) + \int_{M^4} P_4^*(g)(1)\varphi\,\mathrm{vol}(g). \end{aligned}$$

But $P_4^*(1) = P_4(1) = 0$. This yields the assertion.

Although the conformal transformation laws of Q_2 and $Q_{4,n}$ are more complicated in the respective dimensions $n \neq 2$ and $n \neq 4$, (1.2.25) and Theorem 1.2.2 admit the following generalizations to general dimensions.

Theorem 1.2.3. *On manifolds (M,g) of dimension $n \geq 2$, the Yamabe operator*

$$P_{2,n}(g) = \Delta_g - \left(\frac{n}{2}-1\right)Q_2(g) = \Delta_g - \frac{n-2}{2(n-1)}\mathrm{scal(g)} \tag{1.2.26}$$

is conformally covariant, i.e.,

$$e^{(\frac{n}{2}+1)\varphi} P_{2,n}(e^{2\varphi}g)(u) = P_{2,n}(g)(e^{(\frac{n}{2}-1)\varphi}u) \tag{1.2.27}$$

for all $u, \varphi \in C^\infty(M)$.

Proof. A direct calculation shows that

$$e^{(\frac{n}{2}+1)\varphi} \Delta_{e^{2\varphi}g}(e^{-(\frac{n}{2}-1)\varphi}u) = \Delta u - \left(\frac{n}{2}-1\right)\Delta(\varphi)u - \left(\frac{n}{2}-1\right)^2 |\operatorname{grad}(\varphi)|^2 u.$$

On the other hand, (1.2.16) states that

$$e^{2\varphi} \mathsf{J}(e^{2\varphi}g) = \mathsf{J} - \Delta(\varphi) - \left(\frac{n}{2}-1\right)|\operatorname{grad}(\varphi)|^2.$$

These formulas imply the assertion. □

Theorem 1.2.4. *On manifolds* (M, g) *of dimension* $n \geq 3$, *the Paneitz operator*

$$P_{4,n}(g) = \Delta^2 + \delta((n-2)\mathsf{J}g - 4\mathsf{P})d + \left(\frac{n}{2}-2\right) Q_{4,n} \tag{1.2.28}$$

is conformally covariant, i.e.,

$$e^{(\frac{n}{2}+2)\varphi} P_4(e^{2\varphi}g)(u) = P_4(g)(e^{(\frac{n}{2}-2)\varphi}u) \tag{1.2.29}$$

for all $u, \varphi \in C^\infty(M)$.

Of course, $P_{4,n}$ can be written also in terms of the trace-free Ricci tensor Ric_0 and the scalar curvature. This yields the more explicit, but less illuminating formula

$$\Delta^2 + \delta\left(a_n \mathrm{scal} g + \mathrm{b_n} \, \mathrm{Ric}_0\right) d + \left(\frac{n}{2}-2\right) Q_{4,n}$$

with

$$a_n = \frac{n^2-2n-4}{2n(n-1)}, \quad b_n = -\frac{4}{n-2}.$$

We emphasize the important fact that for $n \neq 4$ the constant term of $P_{4,n}$ is a non-trivial multiple of $Q_{4,n}$. However, the critical Q-curvature Q_4 is *not visible* in the critical Paneitz operator P_4 due to the coefficient $\frac{n}{2} - 2$.

Moreover, the situation is analogous to that for Q_2: in (1.2.26), the coefficient $\frac{n}{2} - 1$ hides Q_2 for $n = 2$.

Theorem 1.2.4 was discovered by Paneitz in 1983. Although his work remained unpublished until recently [Pa83], it had a major influence on the subject. In dimension $n = 4$, the operator P_4 also appeared in the independent works [ES85] and [R84].

Next, we give a *proof* of the conformal covariance of the critical Paneitz operator P_4. It rests on the conformal covariance of the Yamabe operator in dimension four (Theorem 1.2.3). In the first step, we write P_4 in the form

$$P_4 = (P_2^2)^0 - 4\delta(\mathsf{P}d) = (P_2^2)^0 - 4\mathcal{T}$$

with $\mathcal{T} = \delta(\mathsf{P}d)$. Now the relation

$$e^{3\varphi} P_2(e^{2\varphi} g) = P_2(g) e^{\varphi}$$

(see (1.2.27)) implies

$$e^{4\varphi} \hat{P}_2^2 = e^{\varphi} P_2 e^{-2\varphi} P_2 e^{\varphi}.$$

Here P_2 and $\hat{P}_2$ are the respective Yamabe operators for g and $\hat{g} = e^{2\varphi} g$. Hence

$$\begin{aligned} (d/dt)|_0 (e^{4t\varphi} P_2^2(e^{2t\varphi} g)) &= \varphi P_2^2 - 2P_2 \varphi P_2 + P_2^2 \varphi \\ &= [P_2, [P_2, \varphi]]. \end{aligned}$$

It follows that

$$(d/dt)|_0 \left(e^{4t\varphi} (P_2^2)^0 (e^{2t\varphi} g)\right) = [P_2, [P_2, \varphi]]^0 = [\Delta, [\Delta, \varphi]]^0.$$

Now a calculation shows that

$$[\Delta, [\Delta, \varphi]]^0 u = 4(\mathrm{Hess}(u), \mathrm{Hess}(\varphi)) + \text{ first-order terms.} \tag{1.2.30}$$

Next, using

$$\mathcal{T}(u) = -(\mathsf{P}, \mathrm{Hess}(u)) + \text{ first-order terms}$$

and (1.2.15), we find

$$(d/dt)|_0 \left(e^{4t\varphi} \mathcal{T}(e^{2t\varphi} g) u\right) = (\mathrm{Hess}(u), \mathrm{Hess}(\varphi)) + \text{ first-order terms.} \tag{1.2.31}$$

Now (1.2.30) and (1.2.31) show that the conformal variation

$$(d/dt)|_0 \left(e^{4t\varphi} P_4(e^{2t\varphi} g) u\right) \tag{1.2.32}$$

is a differential operator of first order. By the self-adjointness of P_4, this operator is self-adjoint, too. Moreover, it has vanishing constant term. Thus, the conformal variation (1.2.32) of P_4 vanishes, and we find

$$\begin{aligned} e^{4\varphi} P_4(e^{2\varphi} g) - P_4(g) &= \int_0^1 (d/dt) \left(e^{4t\varphi} P_4(e^{2t\varphi} g)\right) dt \\ &= \int_0^1 e^{4s\varphi} (d/dt)|_0 \left(e^{4t\varphi} P_4(e^{2t\varphi}(e^{2s\varphi} g))\right) ds = 0 \end{aligned}$$

by (1.2.32) for the metrics $e^{2s\varphi} g$. This proves the conformal covariance of P_4.

Now we continue with the description of some applications.

We recall that the *Yamabe problem* asks for given (M, g) to find a metric in the conformal class $[g]$ with constant scalar curvature. We refer to [LP87] for a lucid account. The positive solution of the Yamabe problem motivates us to pose the following problem.

Problem 1.2.1. *Describe the conformal classes which contain a metric with constant* Q_4.

As one result in this direction, we mention the following theorem [DM08].

Theorem 1.2.5. *Assume that the metric* g *on the closed four-manifold* M *satisfies the conditions*

$$\ker P_4(g) = \mathbb{C}$$

and

$$\int_M Q_4(g)\,\mathrm{vol}(g) \neq 16k\pi^2, \ k = 1, 2, \ldots.$$

Then there exists $\varphi \in C^\infty(M)$ *so that* $Q_4(e^{2\varphi}g) = constant$.

The assumptions in Theorem 1.2.5 are conformally invariant and exclude the sphere S^4 with the round metric g_c. In fact, we have seen above that the fundamental identity for Q_4 implies that the value of the integral only depends on the conformal class. Moreover, for (S^4, g_c), we have $Q_4 = 3!$, i.e.,

$$\int_{S^4} Q_4(g_c)\,\mathrm{vol}(g_c) = 3!\frac{8}{3}\pi^2 = 16\pi^2.$$

Theorem 1.2.5 motivates us to ask for a description of the metrics g so that the kernel $\ker P_4(g)$ is trivial, i.e., $\ker P_4(g) = \mathbb{C}$. More generally, the conformal covariance of P_4 implies that the kernel $\ker P_4(g)$ only depends on the conformal class of g. It is of independent interest to study this conformal invariant.

By the fundamental identity for Q_4, the problem of findind a metric with constant Q_4 in the conformal class of g, is equivalent to the problem of solving the equation

$$Q_4(g) + P_4(g)(\varphi) = e^{4\varphi}\bar{Q}_4 \tag{1.2.33}$$

for a constant $\bar{Q}_4$. This problem is *variational*, i.e., (1.2.33) is the Euler-Lagrange equation of a variational problem. More precisely, (1.2.33) characterizes the critical points of the functional

$$I(\varphi) = \int_{M^4} (\varphi P_4(\varphi) + 2Q_4\varphi)\,\mathrm{vol} \\ - \left(\int_{M^4} Q_4\,\mathrm{vol}\right) \log\left(\int_{M^4} \mathrm{vol}(\hat{g}) \Big/ \int_{M^4} \mathrm{vol}(g)\right). \tag{1.2.34}$$

This extends the fact that for closed surfaces the equation

$$Q_2(g) - P_2(g)(\varphi) = e^{2\varphi}\bar{Q}_2$$

is the Euler-Lagrange equation of the functional

$$J(\varphi) = \int_{M^2} (-\varphi P_2(\varphi) + 2Q_2\varphi)\,\mathrm{vol}$$

$$- \left(\int_{M^2} Q_2 \, \mathrm{vol} \right) \log \left(\int_{M^2} \mathrm{vol}(\hat{g}) / \int_{M^2} \mathrm{vol}(g) \right). \quad (1.2.35)$$

In Section 1.3 we shall define the critical Q-curvature Q_n of a Riemannian manifold of even dimension n. It is natural to study Problem 1.2.1 for these curvature invariants, too. We refer to [B03] and [Nd07] for results in this direction.

The functionals J and I appear also in connection with Polyakov's formula for the conformal variation of determinants. In order to describe this connection, we first recall the notion of determinants as introduced by Ray and Singer [RS71].

Let $-\Delta_g$ be the non-negative Laplacian of (M, g). In particular, the Laplacian of the Euclidean metric of $\mathbb{R}^n$ is $\sum_i \partial^2/\partial x_i$. $-\Delta_g$ gives rise to a self-adjoint unbounded operator on $L^2(M, g)$. By the compactness of M, it has a discrete spectrum consisting of real eigenvalues

$$0 = \lambda_0 < \lambda_1 \leq \lambda_2 \leq \cdots$$

of finite multiplicity. We define the spectral zeta function of $-\Delta_g$ by the Dirichlet series

$$\zeta(s) = \sum_{j \geq 1} \frac{1}{\lambda_j^s}.$$

Weyl's law

$$\# \{ j \,|\, \lambda_j \leq t \} \sim c_n t^{\frac{n}{2}}, \; t \to \infty$$

implies that ζ is holomorphic on the half-plane $\mathrm{Re}(s) > n/2$. ζ admits a meromorphic continuation to the complex plane $\mathbb{C}$. This continuation is regular at $s = 0$ and we define the *zeta-regularized determinant* of $-\Delta$ by

$$\det(-\Delta) = \exp(-\zeta'(0)), \quad (1.2.36)$$

where the prime denotes the complex derivative.

In a similar way, one defines the (zeta-regularized) determinant of more general operators D. If the spectrum of D contains (finitely many) negative eigenvalues, one sets

$$\det(D) = (-1)^{\#\{j | \lambda_j < 0\}} \exp(-\zeta'(0)),$$

where

$$\zeta(s) = \sum_{\lambda_j \neq 0} \frac{1}{|\lambda_j|^s}.$$

We recall that on surfaces, $P_2 = \Delta$ and $Q_2 = K$. The following formula was discovered by Polyakov [Po81].

Theorem 1.2.6. *For volume-preserving conformal changes $g \mapsto e^{2\varphi} g$ on closed surfaces (M^2, g),*

$$\log \left(\frac{\det(-P_2(e^{2\varphi} g))}{\det(-P_2(g))} \right) = -\frac{1}{12\pi} \int_{M^2} [-\varphi P_2(g) \varphi + 2 Q_2(g) \varphi] \, \mathrm{vol}(g). \quad (1.2.37)$$

For conformal changes which do not preserve the volume, the quotient

$$\log\left(\int_{M^2}\operatorname{vol}(\hat{g})/\int_{M^2}\operatorname{vol}(g)\right)$$

does not vanish and yields an additional contribution in (1.2.37). Detailed proofs can be found in [Br93], [C04]. For a discussion from the point of view of string theory we refer to paragraph 3.5 in the lectures of Eric D'Hoker in [D99].

One of the remarkable aspects of (1.2.37) is that the change of the transcendental determinant within conformal classes is described by *local* terms: Gauß-curvature.

We note that the Gauß curvature prescription equation (1.2.12) implies that

$$Q_2(g)\operatorname{vol}(g) + Q_2(\hat{g})\operatorname{vol}(\hat{g}) = (Q_2(g) + (Q_2(g) - P_2(g)(\varphi)))\operatorname{vol}(g)$$

for $\hat{g} = e^{2\varphi}g$. Hence

$$\int_{M^2}[-\varphi P_2(g)(\varphi) + 2Q_2(g)\varphi]\operatorname{vol} = \int_{M^2}\varphi\left[Q_2(g)\operatorname{vol}(g) + Q_2(\hat{g})\operatorname{vol}(\hat{g})\right]. \quad (1.2.38)$$

In particular, the integral on the right-hand side of (1.2.37) is only determined by the Gauß curvature. In the physical literature, the functional

$$S(\varphi,g) = \frac{1}{12\pi}\int_{M^2}\left[\frac{1}{2}|\operatorname{grad}_g(\varphi)|^2 + K(g)\varphi\right]\operatorname{vol}(g)$$

is called the *Liouville action* (see [D99]). From the point of view of (1.2.38), it is more natural to consider the functional

$$\mathcal{H}:(\hat{g},g)\mapsto\int_{M^2}\varphi\left[Q_2(g)\operatorname{vol}(g) + Q_2(\hat{g})\operatorname{vol}(\hat{g})\right]$$

of two arguments. $\mathcal{H}$ is alternating. Moreover, the Gauß-curvature prescription equation implies the *cocycle relation*

$$\mathcal{H}(e^{2(\psi+\varphi)}g,g) = \mathcal{H}(e^{2(\psi+\varphi)}g,e^{2\varphi}g) + \mathcal{H}(e^{2\varphi}g,g). \quad (1.2.39)$$

In order to prove (1.2.39), it will be convenient to introduce the Lagrangian

$$\mathcal{L}(\hat{g},g) = \varphi\left[Q_2(g)\operatorname{vol}(g) + Q_2(\hat{g})\operatorname{vol}(\hat{g})\right],\ \hat{g} = e^{2\varphi}g.$$

Then $\mathcal{H}(\hat{g},g) = \int_M\mathcal{L}(\hat{g},g)$. Now an application of the Gauß-curvature prescription equation and the conformal covariance of P_2 yields

$$\begin{aligned}\mathcal{L}(e^{2(\varphi+\psi)}g,e^{2\varphi}g) &= \psi\left[2Q_2(e^{2\varphi}g) - P_2(e^{2\varphi}g)(\psi)\right]\operatorname{vol}(e^{2\varphi}g)\\ &= \psi\left[2Q_2(g) - 2P_2(g)(\varphi) - P_2(g)(\psi)\right]\operatorname{vol}(g)\end{aligned}$$

and

$$\mathcal{L}(e^{2\varphi}g,g) = \varphi\,[2Q_2(g) - P_2(g)(\varphi)]\,\mathrm{vol}(g).$$

Hence

$$\begin{aligned}\mathcal{L}(e^{2(\varphi+\psi)}g, e^{2\varphi}g) + \mathcal{L}(e^{2\varphi}g,g)\\ = \mathcal{L}(e^{2(\varphi+\psi)}g,g) + [\varphi P_2(g)(\psi) - \psi P_2(g)(\varphi)])\,\mathrm{vol}(g).\end{aligned}$$

But $\varphi P_2(g)(\psi) - \psi P_2(g)(\varphi) = \delta(\alpha(\varphi,\psi))$ with $\alpha(\varphi,\psi) = \psi d\varphi - \varphi d\psi \in \Omega^1(M)$. This proves (1.2.39). It plays a fundamental role in [KS06].

Formula (1.2.37) naturally applies to the study of the extremal properties of the determinant. In particular, each conformal class on a surface admits a metric which extremizes the determinant. This yields a proof of the Uniformization Theorem [OPS88a], [OPS88b].

Theorem 1.2.6 admits far-reaching generalizations to operators on manifolds of higher dimensions. Let D be a *natural* formally self-adjoint linear differential operator on M with positive principal symbol. Here an operator D is called natural if it is defined in terms of a metric g by using a number of standard operations (see Section 1.3 for a more precise definition). In other words, a natural operator is a rule to associate an operator $D(g)$ to any metric g. Moreover, we assume that D is *conformally covariant* in the sense that there exist constants $a, b \in \mathbb{R}$ so that

$$e^{a\varphi} \circ D(e^{2\varphi}g) = D(g) \circ e^{b\varphi}$$

for all $\varphi \in C^\infty(M)$. For the precise meaning of the technical assumptions we refer to [Br93]. Typical examples for D are the Yamabe operator and the Paneitz operator. The following result was proved in [BO91b].

Theorem 1.2.7. *Let D be as above. Assume that* $\ker D(g) = 0$. *Then, for volume-preserving conformal changes $g \mapsto \hat{g} = e^{2\varphi}g$ on the closed four-manifold (M,g),*

$$\begin{aligned}\log\left(\frac{\det D(\hat{g})}{\det D(g)}\right) &= a\int_{M^4} [\varphi P_4(\varphi) + 2Q_4\varphi]\,\mathrm{vol}\\ &+ b\int_{M^4} \varphi|\mathsf{W}|^2\,\mathrm{vol} + c\left(\int_{M^4} \mathsf{J}^2(\hat{g})\,\mathrm{vol}(\hat{g}) - \int_{M^4}\mathsf{J}^2(g)\,\mathrm{vol}(g)\right) \quad (1.2.40)\end{aligned}$$

for certain constants a, b, c which depend on the operator D but not on the metric.

Remark 1.2.1. The constants a, b, c in Theorem 1.2.7 arise through the identification of the constant term in the expansion of the trace of the heat kernel of D as a linear combination of Q_4, $|\mathsf{W}|^2$ and $\Delta\mathsf{J}$. In particular, for $D = -P_2$ [BO91b],

$$(a,b,c) \sim (-1, 1, -2)$$

and for $D = P_4$ [Br96],

$$(a,b,c) \sim (-14, -1, 32).$$

For conformal changes which do not preserve the volume, the quotient

$$\log\left(\int_{M^4} \operatorname{vol}(\hat{g}) / \int_{M^2} \operatorname{vol}(g)\right)$$

does not vanish and yields additional contributions in the corresponding version of (1.2.40).

For the conformal class of the round sphere S^4, the contribution of W in (1.2.40) vanishes, and by Remark 1.2.1, the coefficients a and c for the Yamabe operator have the same sign. This leads to the following application of Theorem 1.2.7.

Theorem 1.2.8 ([BCY92]). *On the round sphere (S^4, g_c), we consider the determinant of $-P_2(g)$ for volume-preserving conformal changes of g_c. It is minimal iff g is the pull-back of g_c by a conformal diffeomorphism.*

Theorem 1.2.8 is the four-dimensional analog of the result of Onofri [O82] that among all metrics g on S^2 of the same volume 4π as the round metric g_c the determinant of $-P_2(g) = -\Delta_g$ is *maximal* if g is the pull-back of g_0 by a diffeomorphism.

Similarly as for surfaces, the fundamental identity for Q_4 implies that the first term on the right-hand side of the variational formula (1.2.40) can be given a more natural formulation as follows. In fact, Theorem 1.2.1 shows that

$$Q_4(g)\operatorname{vol}(g) + Q_4(\hat{g})\operatorname{vol}(\hat{g}) = [Q_4(g) + (Q_4(g) + P_4(g)(\varphi))]\operatorname{vol}(g)$$

for $\hat{g} = e^{2\varphi}g$. Hence

$$\int_{M^4} [\varphi P_4(g)(\varphi) + 2Q_4(g)\varphi]\operatorname{vol}(g) = \int_{M^4} \varphi\,[Q_4(g)\operatorname{vol}(g) + Q_4(\hat{g})\operatorname{vol}(\hat{g})]\,.$$

Moreover, the two-variable function

$$\mathcal{H}(\hat{g}, g) = \int_{M^4} \varphi\,[Q_4(g)\operatorname{vol}(g) + Q_4(\hat{g})\operatorname{vol}(\hat{g})]\,,\ \hat{g} = e^{2\varphi}g$$

on conformal classes satisfies the cocycle relation (1.2.39). Its proof is analogous to that for $\mathcal{H}$ on closed surfaces. It rests only on the conformal transformation law of Q_4 and the self-adjointness of P_4. Moreover, it generalizes to arbitrary even dimensions in terms of the corresponding critical Q-curvatures. For a detailed discussion of these cocycles we refer to [Br05].

For the round sphere (S^4, g_c), a crucial ingredient in the proof of Theorem 1.2.8 is the inequality

$$\int_{S^4} \varphi\left[Q_4(g_c)\operatorname{vol}(g_c) + Q_4(e^{2\varphi}g_c)\operatorname{vol}(e^{2\varphi}g_c)\right] \geq 0$$

for volume-preserving conformal changes of g_c. This inequality is the special case $n = 4$ of an inequality which is equivalent to an inequality due to Beckner [Be93]. For more details we refer to chapter 3 of [Br93] and [Br05].

Next, we describe an important connection between the Q-curvature Q_4 and the Gauß-Bonnet formula. We first note that in dimension $n = 2$, the total Q-curvature

$$\int_{M^2} Q_2 \,\mathrm{vol}$$

of a closed surface (M^2, g) equals $2\pi\chi(M)$ by the Gauß-Bonnet formula. Similarly, we have seen that the total Q-curvature

$$\int_{M^4} Q_4 \,\mathrm{vol} \tag{1.2.41}$$

is a conformal invariant of a four-manifold (M^4, g). In order to evaluate the integral, we apply the Chern-Gauß-Bonnet theorem

$$\int_{M^4} \mathrm{Pf}_4 = (2\pi)^2 \chi(M). \tag{1.2.42}$$

Here $\mathrm{Pf}_4 \in \Omega^4(M)$ is the Pfaffian form. Identifying Pf_4 with a multiple of the Riemannian volume form, we find

$$\mathrm{Pf}_4 = (\mathsf{J}^2 - |\mathsf{P}|^2) + \frac{1}{8}|\mathsf{W}|^2. \tag{1.2.43}$$

It follows that

$$Q_4 = 2\,\mathrm{Pf}_4 - \frac{1}{4}|\mathsf{W}|^2 - \Delta \mathsf{J}. \tag{1.2.44}$$

The latter formula implies that the total Q-curvature (1.2.41) equals

$$8\pi^2\chi(M) - \frac{1}{4}\int_{M^4} |\mathsf{W}|^2 \,\mathrm{vol},$$

i.e., is a linear combination of the *topological* invariant $\chi(M)$ and the *conformal* invariant $\mathcal{W}$.

The relation (1.2.43) also implies that the total integral of the curvature quantity $\mathsf{J}^2 - |\mathsf{P}|^2$ is conformally invariant (in dimension 4). Viaclovsky [V00] discovered that in dimension $n \neq 4$, the corresponding integral defines a functional with an interesting Euler-Lagrange equation. Thus, we consider the functional

$$\mathcal{V}_4 = \int_M (\mathsf{J}^2 - |\mathsf{P}|^2) \,\mathrm{vol} \Big/ \left(\int_M \mathrm{vol} \right)^{\frac{n-4}{n}} \tag{1.2.45}$$

on closed Riemannian manifolds M^n of dimension $n \geq 3$.

Theorem 1.2.9. *For $n \neq 4$ and a fixed metric g, the conformal variations*

$$(d/dt)|_0 \left(\mathcal{V}_4(e^{2t\varphi}g)\right)$$

vanish for all $\varphi \in C^\infty(M)$ iff for the metric g the quantity

$$\mathsf{J}^2 - |\mathsf{P}|^2 \tag{1.2.46}$$

is constant.

Proof. The transformation rules (1.2.15) and (1.2.16) imply that the functional $\mathcal{V}_4^0 = \int_M (\mathsf{J}^2 - |\mathsf{P}|^2)\,\mathrm{vol}$ satisfies

$$\mathcal{V}_4^0(e^{2t\varphi}g) = \int_M e^{(n-4)t\varphi}\left(\mathsf{J}^2 - |\mathsf{P}|^2 - 2t\mathsf{J}\Delta\varphi + 2t(\mathsf{P}, \mathrm{Hess}(\varphi))\right)\mathrm{vol}(g),$$

up to terms which are at least quadratic in t. Thus, we find the variational formula

$$\begin{aligned}&(d/dt)|_0(\mathcal{V}_4^0(e^{2t\varphi}g))\\ &\qquad = (n-4)\int_M \varphi(\mathsf{J}^2 - |\mathsf{P}|^2)\,\mathrm{vol}(g) + 2\int_M (-\mathsf{J}\Delta\varphi + (\mathsf{P}, \mathrm{Hess}(\varphi))\,\mathrm{vol}(g).\end{aligned}$$

By partial integration, the second integral equals

$$\int_M ((d\mathsf{J}, d\varphi) - (\delta(\mathsf{P}), d\varphi))\,\mathrm{vol}, \tag{1.2.47}$$

where the divergence $\delta(\mathsf{P}) \in \Omega^1(M)$ is defined by $\sum_i \nabla_i(\mathsf{P})(e_i, \cdot)$. Now the relation $\delta(\mathsf{P}) = d\mathsf{J}$ shows that the integral (1.2.47) vanishes. Hence

$$\begin{aligned}&(d/dt)_0(\mathcal{V}_4(e^{2t\varphi}g))\\ &= (n-4)\int_M \varphi(\mathsf{J}^2 - |\mathsf{P}|^2)\,\mathrm{vol}(g)/\left(\int_M \mathrm{vol}(g)\right)^{\frac{n-4}{n}} - (n-4)\mathcal{V}_4(g)\left(\frac{\int_M \varphi\,\mathrm{vol}(g)}{\int_M \mathrm{vol}(g)}\right),\end{aligned}$$

and it follows that the conformal variation vanishes iff

$$\int_M \varphi(\mathsf{J}^2 - |\mathsf{P}|^2)\,\mathrm{vol}(g) = \int_M (\mathsf{J}^2 - |\mathsf{P}|^2)\,\mathrm{vol}(g)\left(\frac{\int_M \varphi\,\mathrm{vol}(g)}{\int_M \mathrm{vol}(g)}\right).$$

The proof is complete. □

One can regard P as an endomorphism of TM by setting

$$g(\mathsf{P}(X), Y) = \mathsf{P}(X, Y),\ X, Y \in \mathfrak{X}(M).$$

This endomorphism induces an endomorphism $\wedge^2(\mathsf{P})$ of the bundle $\wedge^2(TM)$. In these terms, we have

$$\mathsf{J}^2 - |\mathsf{P}|^2 = 2\,\mathrm{tr}\wedge^2(\mathsf{P}). \tag{1.2.48}$$

The latter identity follows by applying the obvious algebraic relation $(\sum_i x_i)^2 - \sum_i x_i^2 = 2\sum_{i<j} x_i x_j$ to the eigenvalues of the endomorphism P. The interpretation (1.2.48) of $\mathsf{J}^2 - |\mathsf{P}|^2$ suggests to consider the more general quantities $\sigma_p = \operatorname{tr} \wedge^p(\mathsf{P})$ and the associated functionals

$$\mathcal{V}_{2p} = \int_M \operatorname{tr} \wedge^p(\mathsf{P}) \operatorname{vol}.$$

This idea has led to a substantial literature in conformal geometric analysis. For a recent review see [V06]. For locally conformally flat metrics, the quantities σ_p are closely connected to the holographic coefficients v_{2j} of Definition 1.5.1:

$$\sigma_p = (-2)^p v_{2p}.$$

For general metrics, the holographic coefficients give rise to a class of new interesting variational problems [CF08].

The identity

$$\int_{M^4} Q_4 \operatorname{vol} + \frac{1}{4}\int_{M^4} |\mathsf{W}|^2 \operatorname{vol} = 8\pi^2 \chi(M^4)$$

relates *three* global conformal invariants. The relative size of these invariants can be used to characterize some conformal metrics. The following theorem is a remarkable result in this direction. Let

$$Y(M,g) = \inf_{\hat{g}\in[g]} \left(\int_{M^4} \operatorname{scal}(\hat{g}) \operatorname{vol}(\hat{g})\right) \Big/ \left(\int_{M^4} \operatorname{vol}(\hat{g})\right)^{\frac{n-2}{n}}$$

be the Yamabe constant.

Theorem 1.2.10 ([CGY03]). *Let (M,g) be a closed four-manifold with positive Yamabe constant $Y(M,g)$. If*

$$\int_M Q_4(g) \operatorname{vol}(g) > \frac{1}{4}\int_M |\mathsf{W}(g)|^2 \operatorname{vol}(g),$$

then M is diffeomorphic to the round sphere S^4 or the real projective space $\mathbb{R}P^4$. If M is not diffeomorphic to these spaces and

$$\int_M Q_4(g) \operatorname{vol}(g) = \frac{1}{4}\int_M |\mathsf{W}(g)|^2 \operatorname{vol}(g),$$

then M is conformally equivalent to $\mathbb{C}P^2$ with the Fubini-Study metric or a quotient $\Gamma\backslash(S^3 \times S^1)$ with the product metric.

For more information on the interactions of partial differential equations with conformal geometry see [CY02], [C05] and [Gu09].

1.3 GJMS-operators and Branson's Q-curvatures

On general manifolds, the pairs (P_2, Q_2) and (P_4, Q_4) considered in Section 1.2 are only the first two elements in a sequence of higher-order constructions. The relevant high-order operators were constructed in the celebrated work [GJMS92]. It is common to refer to these operators as the GJMS-operators. Shortly after [GJMS92] Branson [Br93] used the GJMS-operators to construct a sequence of higher-order analogs of Q_2 and Q_4. Now these curvature quantities are called Branson's Q-curvatures.

In the present section, we describe the basic constructions of GJMS-operators and Branson's Q-curvatures.

We start with the description of the conformally covariant powers of the Laplacian. As in Section 1.2, we shall use the convention that $-\Delta \geq 0$. The following result extends the construction of the operators P_2 and P_4 in Section 1.2.

Theorem 1.3.1. *Let the natural numbers $n \geq 3$ and $N \geq 1$ be subject to the condition $1 \leq N \leq n/2$ if n is even. Then on any Riemannian manifold (M, g) of dimension n, there exists a natural differential operator P_{2N} on $C^\infty(M)$ of the form*

$$P_{2N}(g) = \Delta_g^N + LOT \tag{1.3.1}$$

so that

$$e^{(\frac{n}{2}+N)\varphi} \circ P_{2N}(e^{2\varphi} g) \circ e^{-(\frac{n}{2}-N)\varphi} = P_{2N}(g) \tag{1.3.2}$$

for all $\varphi \in C^\infty(M)$. On the flat space $(\mathbb{R}^n, g_c)$, the operator P_{2N} is Δ_c^N.

Theorem 1.3.1 extends to the case of metrics of arbitrary signature. The operators in Theorem 1.3.1 will be referred to as the GJMS-operators. For even n, P_n will be called the *critical* GJMS-operator, and P_{2N} for $2N < n$ are called *subcritical.* In these terms, the Yamabe and Paneitz operators are critical for surfaces and four-manifolds, respectively.

Before we outline the idea of the proof of Theorem 1.3.1, we add a series of comments. In Theorem 1.3.1, the property of being *natural* means that the operators are given by universal formulas in the metric g, its inverse, the Levi-Civita connection and the curvature of g using tensor product and contraction. Thus to give a natural operator means to give a rule which associates an operator to any manifold with a metric.

As formulated, Theorem 1.3.1 is an existence result. Its proof actually establishes the existence of the operators P_{2N} through a construction. This also offers the possibility to derive explicit formulas, at least in principle. In fact, finding explicit formulas for GJMS-operators is a challenge. It also involves finding the most natural form to state the results. For recent progress in this direction we refer to [J09a]. The problem will be touched again in Section 1.6.

The GJMS-operators arise by adding geometrically defined lower-order terms to corresponding powers of the Laplacian. By this reason, it is also natural to refer to them as conformally covariant powers of the Laplacian. On the other hand, Theorem 1.3.1 leaves open the question of determining the structure of *all* conformally covariant powers of the Laplacian. In fact, they are not uniquely determined by the requirements in Theorem 1.3.1. As an easy example we note that, for any constant a, the operator

$$P_4 + a|\mathsf{W}|^2$$

has the same properties as P_4.

On manifolds of odd dimensions, Theorem 1.3.1 yields an infinite sequence of operators $P_2, P_4, P_6, \dots$. However, on manifolds of even dimension, Theorem 1.3.1 yields only a finite sequence which terminates at the critical operator. In the proof, the restriction to the range $2N \leq n$ is caused by the obstructed smoothness of the Fefferman-Graham ambient metric. On the other hand, there is *no* conformally covariant power of the Laplacian, the order of which exceeds the (even) dimension of the underlying space. For cubes on four manifolds this was proved in [G92]. For the general case see [GH04].

Under additional restrictions on the conformal classes, the obstructions may vanish. In such cases, the construction of [GJMS92] yields operators in the range $2N > n$.

Now we outline the main ideas of the proof of Theorem 1.3.1. For a given Riemannian manifold (M,g) of dimension n, we consider the ray bundle

$$\mathcal{G}_M = \{(x, tg_x),\ t > 0\} \subset S^2T^*M$$

with the canonical projection $\pi : \mathcal{G}_M \to M$. Sections of $\mathcal{G}_M$ are metrics in the conformal class $[g]$. $\mathcal{G}_M$ should be regarded as a substitute of the upper light-cone

$$C^+ = \left\{(x_0, x') \in \mathbb{R}^{(1,n+1)} \,|\, b(x) = 0,\ x_0 > 0\right\}$$

with the projection $C^+ \ni (x_0, x') \mapsto x'/x_0 \in S^n$ together with the round metric g_c on S^n. In this case, we identify $(x_0, x') \in C^+$ with $(y, x_0^2(g_c)_y) \in \mathcal{G}_{S^n}$, where $y = x'/x_0$. A natural $\mathbb{R}^+$-action on the space $\mathcal{G}_M$ is defined by

$$\delta_s : (x, tg_x) \mapsto (x, s^2tg_x),\ s > 0.$$

It substitutes the $\mathbb{R}^+$-action $x \mapsto sx$ on C^+. δ_s gives rise to a notion of homogeneous functions on $\mathcal{G}_M$. More precisely, for $\lambda \in \mathbb{C}$, we define

$$\mathcal{E}_M(\lambda) = \left\{u \in C^\infty(\mathcal{G}_M) \,|\, \delta_s^*(u) = s^\lambda u,\ s \in \mathbb{R}^+\right\}. \tag{1.3.3}$$

More geometrically, elements of $\mathcal{E}_M(\lambda)$ can be identified with sections of the line bundle $\mathcal{L}_\lambda$ on M which is associated to the $\mathbb{R}^+$-bundle $\mathcal{G}_M$ by the character s^λ of $\mathbb{R}^+$. The choice of a metric in the conformal class $[g]$ induces a trivialization of $\mathcal{L}_\lambda$ and an isomorphism

$$T_g(\lambda) : C^\infty(M) \to \mathcal{E}_M(\lambda), \quad u \mapsto v(x, tg_x) = t^{\lambda/2}u(x).$$

It follows that

$$T_{e^{2\varphi}g}(\lambda) = T_g(\lambda) \circ e^{-\lambda\varphi}. \tag{1.3.4}$$

Next, there is a tautological symmetric bilinear form $\mathbf{g}$ on $T(\mathcal{G}_M)$ so that $\delta_s^*(\mathbf{g}) = s^2\mathbf{g}$. It substitutes the restriction of the Lorentzian metric on $\mathbb{R}^{1,n+1}$ to C^+. $\mathbf{g}$ is defined by

$$\mathbf{g}_{(x,tg_x)}(Y,Z) = (tg_x)(\pi_*(Y),\pi_*(Z)) \quad \text{for} \quad Y,Z \in T_{(x,tg_x)}(\mathcal{G}_M). \tag{1.3.5}$$

Now, Fefferman and Graham [FG85] discovered a natural substitute of the Lorentzian metric on $\mathbb{R}^{1,n+1}$ in the general setting. In fact, on the thickening

$$\tilde{\mathcal{G}}_M = \mathcal{G}_M \times (-1,1)$$

of $\mathcal{G}_M$, we consider metrics $\tilde{g}$ of signature $(1,n+1)$ with the properties

- $i^*(\tilde{g}) = \mathbf{g}$,
- $\tilde{\delta}_s^*(\tilde{g}) = s^2\tilde{g}$,
- $\operatorname{Ric}(\tilde{g}) = 0$ along $\mathcal{G}_M$.

Here $i : \mathcal{G}_M \hookrightarrow \tilde{G}_M$ denotes the embedding $m \mapsto (m,0)$ and the $\mathbb{R}^+$-action $\tilde{\delta}_s$ naturally extends δ_s. Note that these properties are preserved under $\mathbb{R}^+$-equivariant diffeomorphisms which restrict to the identity on $\mathcal{G}_M$. The third condition requires some additional comments. If the dimension of M is *odd*, there are metrics $\tilde{g}$ for which the Ricci tensor vanishes of infinite order on $\mathcal{G}_M$. Moreover, these conditions uniquely determine $\tilde{g}$, up to diffeomorphisms which restrict to the identity on $\mathcal{G}_M$. For *even* dimension n of M, the situation is more complicated. In this case, it is possible to satisfy the vanishing of $\operatorname{Ric}(\tilde{g})$ on $\mathcal{G}_M$ up to order $\frac{n}{2}-1$ (for the tangential components). Moreover, up to the addition of terms which vanish of order $\frac{n}{2}$ and up to $\mathbb{R}^+$-equivariant diffeomorphisms, the metric $\tilde{g}$ is uniquely determined. The existence of $\tilde{g}$ with $\operatorname{Ric}(\tilde{g})$ vanishing to order $\frac{n}{2}$ is obstructed by the Fefferman-Graham *obstruction tensor.* The metrics $\tilde{g}$ are called Fefferman-Graham *ambient metrics.* The full details of these constructions are given in [FG07].

Now the GJMS-operators P_{2N} are derived from the powers of the Laplacian of the ambient metric as follows. First of all, for all $\lambda \in \mathbb{C}$, the operator $\Delta_{\tilde{g}}^N$ induces an operator

$$\Delta_{\tilde{g}}^N : \tilde{\mathcal{E}}_M(\lambda) \to \tilde{\mathcal{E}}_M(\lambda-2N)$$

on $\tilde{\delta}_s$-homogeneous functions on $\tilde{\mathcal{G}}_M$. Now the key observation is that for $\lambda = -\frac{n}{2}+N$, the operator $\Delta_{\tilde{g}}^N$ descends to an operator on homogeneous functions on $\mathcal{G}_M$. More precisely, the composition

$$P_{2N} : \mathcal{E}_M\left(-\frac{n}{2}+N\right) \ni u \xrightarrow{\text{extension}} \tilde{u} \mapsto i^*\Delta_{\tilde{g}}^N(\tilde{u}) \in \mathcal{E}_M\left(-\frac{n}{2}-N\right)$$

is well-defined, i.e., does not depend on the choice of the extension $\tilde{u}$ of u into a neighbourhood of $\mathcal{G}_M$ in $\tilde{\mathcal{G}}_M$. By the uniqueness of $\tilde{g}$ up to diffeomorphisms which fix $\mathcal{G}_M$, the resulting operator P_{2N} does not depend on the choice of the ambient metric. Finally, by composition with the isomorphisms $T_g(\lambda)$, P_{2N} induces operators

$$P_{2N}(g) : C^\infty(M) \to C^\infty(M),$$

and (1.3.4) implies that these operators satisfy the relations

$$P_{2N}(e^{2\varphi}g) = e^{(-\frac{n}{2}-N)\varphi} \circ P_{2N}(g) \circ e^{-(-\frac{n}{2}+N)\varphi}.$$

This completes the proof.

Now let n be even and P_{2N} $(2N \leq n)$ as in Theorem 1.3.1. Then Branson [Br93], [Br95] showed that P_{2N} is of the form

$$P_{2N} = \delta S_{2N-2}d + (-1)^N \left(\frac{n}{2} - N\right) Q_{2N} \tag{1.3.6}$$

for a natural differential operator $S_{2N-2} : \Omega^1(M) \to \Omega^1(M)$ and a local scalar Riemannian invariant Q_{2N}. In particular, the critical GJMS-operator P_n has vanishing constant term. Hence (1.3.6) defines Q_{2N} only if $2N < n$. A closer analysis of the construction of P_{2N} shows that Q_{2N} is a rational function in n without a pole in $n = 2N$. Thus, Q_n is well-defined, too. The formulas for Q_4 given in Section 1.2 may serve as an illustration. [1]

The scalar Riemannian curvature invariants Q_{2N} are of order $2N$, i.e., their definition involves $2N$ derivatives of the metric.

Definition 1.3.1 (Q-curvature). On manifolds of even dimension n, the scalar Riemannian curvature invariants Q_{2N}, $2N \leq n$ are called Branson's Q-curvatures. Moreover, the quantity Q_n is called the critical Q-curvature, and the Q-curvatures $Q_2, \dots, Q_{n-2}$ are called subcritical.

Similarly, for odd n, the constant terms of the GJMS-operators give rise to an infinite sequence $Q_2, Q_4, Q_6, \dots$ of Q-curvatures. In the following, we shall be interested mostly in manifolds of even dimension.

The critical GJMS-operators and the critical Q-curvatures play a distinguished role. In fact, similarly as for $n = 2$ and $n = 4$, the critical GJMS-operators govern the conformal transformation laws of the critical Q-curvatures. The following result extends (1.2.12) and Theorem 1.2.1.

Theorem 1.3.2 (The fundamental identity). *On Riemannian manifolds (M, g) of even dimension n,*

$$e^{n\varphi}Q_n(e^{2\varphi}g) = Q_n(g) + (-1)^{\frac{n}{2}} P_n(g)(\varphi). \tag{1.3.7}$$

Proof. Under the assumption $2N < n$, we apply (1.3.2) to the function

$$u = e^{(\frac{n}{2}-N)\varphi}.$$

[1] Branson used the opposite sign convention for the Laplacian. The sign $(-1)^N$ in (1.3.6) guarantees that we get the same Q_{2N}.

We find

$$e^{(\frac{n}{2}+N)\varphi} P_{2N}(e^{2\varphi} g)(1) = P_{2N}(g)(e^{(\frac{n}{2}-N)\varphi}).$$

Hence

$$\left(\frac{n}{2} - N\right) e^{(\frac{n}{2}+N)\varphi} Q_{2N}(e^{2\varphi} g)$$
$$= (-1)^N P_{2N}^0(g) \left(e^{(\frac{n}{2}-N)\varphi} - 1\right) + \left(\frac{n}{2} - N\right) e^{(\frac{n}{2}-N)\varphi} Q_{2N}(g).$$

Now we divide the latter equation by $(\frac{n}{2} - N)$ and set $n = 2N$. This gives

$$e^{n\varphi} Q_n(e^{2\varphi} g) = (-1)^{\frac{n}{2}} P_n^0(g)(\varphi) + Q_n(g) = (-1)^{\frac{n}{2}} P_n(g)(\varphi) + Q_n(g).$$

The proof is complete. □

The argument in this proof is often referred to as "analytic continuation in dimension".

As in dimensions $n = 2$ and $n = 4$, the fundamental identity has the following consequence.

Theorem 1.3.3. *On a closed manifold M of even dimension n, the functional*

$$T_n(g) = \int_{M^n} Q_n(g) \operatorname{vol}(g)$$

is conformally invariant, i.e.,

$$T_n(e^{2\varphi} g) = T_n(g) \quad \textit{for all} \quad \varphi \in C^\infty(M).$$

Proof. The fundamental identity (Theorem 1.3.2) yields

$$T_n(e^{2\varphi} g) = \int_M e^{n\varphi} Q_n(e^{2\varphi} g) \operatorname{vol}(g) = \int_M \left[Q_n(g) + (-1)^{\frac{n}{2}} P_n(g)(\varphi)\right] \operatorname{vol}(g).$$

Now

$$\int_M P_n(g)(\varphi) \operatorname{vol}(g) = \int_M \delta S d\varphi \operatorname{vol}(g) = 0$$

by (1.3.6). Hence

$$T_n(e^{2\varphi} g) = T_n(g).$$

The proof is complete. □

Remark 1.3.1. Combining $P_n(1) = 0$ with the fact that P_n is formally self-adjoint, yields a version of the above proof. The self-adjointness of GJMS-operators naturally follows from their role as residues of a scattering operator. This will be discussed in Section 1.4.

The total integral of the critical Q-curvature is an example of a *conformally invariant* scalar-valued functional. Other sources of such functionals are the total integrals of certain heat coefficients of conformally covariant differential operators and the integral

$$\int_{M^n} v_n(g)\,\mathrm{vol}(g)$$

of the holographic anomaly v_n of the renormalized volume of Poincaré-Einstein metrics. For the definitions of these concepts we refer to Section 1.5 (see (1.5.5)). It is a natural problem to ask for the structure of those scalar Riemannian invariants the integrals of which define such functionals. The problem was raised and answered in [DS93]. In fact, Deser and Schwimmer proposed a classification of such invariants. Although this classification is widely used in the physical literature, only recently Alexakis established a rigorous proof of the following decomposition in a series of remarkable works [A06a], [A06b] and [A07].

Theorem 1.3.4. *Let $I(g)$ be a scalar Riemannian invariant of even weight $-n$ with the property that, on any closed manifold M of dimension n, the integral*

$$\int_{M} I(g)\,\mathrm{vol}(g)$$

is conformally invariant. Then, $I(g)$ is a linear combination of the Chern-Gauß-Bonnet integrand, a local conformal invariant of weight $-n$ and the divergence $\delta(\omega)$ of a natural one-form ω.

Here local conformal invariants of weight $-n$ are scalar Riemannian invariants $W(g)$ with the property $e^{n\varphi}W(e^{2\varphi}g) = W(g)$ for all $\varphi \in C^\infty(M)$.

In connection with Theorem 1.3.4 we also mention [BGP95]

Theorem 1.3.5. *The total Q-curvature of locally conformally flat metrics is a multiple of the Euler characteristic of M.*

The latter result motivates us to ask for an explicit decomposition of Q_n as the sum of a multiple of the Chern-Gauß-Bonnet integrand and a divergence. In Section 1.5, we shall describe such a decomposition which implies Theorem 1.3.5 (see Corollary 1.5.3).

Problem 1.3.1. *What is the geometric significance of Q-curvatures? In particular, study the consequences of the existence of metrics with specific Q-curvatures and specific T_n.*

Next, we make explicit the concepts of GJMS-operators and Q-curvatures in the conformally flat case of the round spheres. This will also prepare the identification of GJMS-operators and critical Q-curvature in the scattering operator in Section 1.4.

On the round sphere S^n, the conformal covariance of P_{2N} leads to an interpretation of P_{2N} as an intertwining operator for spherical principal series representations of $G^{n+1} = \mathrm{SO}(1, n+1)^\circ$. In fact, we recall from Section 1.1, that any

$g \in G$ acts on S^n by a conformal diffeomorphism, i.e.,

$$g_*(g_c) = \Phi_g^2 g_c \tag{1.3.8}$$

for a nowhere vanishing $\Phi_g \in C^\infty(S^n)$. Here $g_* = (g^*)^{-1}$ denotes the push-forward induced by g. The conformal factor Φ_g can be given in terms of the matrix g:

$$\Phi_g(x) = \frac{1}{d-(b,x)} \quad \text{for} \quad g = \begin{pmatrix} d & c \\ b^t & A \end{pmatrix}.$$

(1.3.8) yields the relation

$$\Phi_{g_1 g_2} = (g_2)_*(\Phi_{g_1})\Phi_{g_2}. \tag{1.3.9}$$

For the conformal changes $g_c \mapsto \Phi_g^2 g_c$, (1.3.2) implies that

$$\Phi_g^{-\frac{n}{2}-N} \circ P_{2N}(g_c) \circ \Phi_g^{\frac{n}{2}-N} = P_{2N}(g_*(g_c)). \tag{1.3.10}$$

Now we have

$$P_{2N}(g_*(g_c)) = g_* \circ P_{2N}(g_c) \circ g^*$$

by the naturality of P_{2N}. Therefore, (1.3.10) can be written in the form

$$\Phi_g^{-\frac{n}{2}-N} \circ P_{2N}(g_c) \circ \Phi_g^{\frac{n}{2}-N} = g_* \circ P_{2N}(g_c) \circ g^*,$$

i.e.,

$$P_{2N}(g_c) \circ \left(\Phi_g^{\frac{n}{2}-N} g_*\right) = \left(\Phi_g^{\frac{n}{2}+N} g_*\right) \circ P_{2N}(g_c). \tag{1.3.11}$$

(1.3.9) shows that for any $\lambda \in \mathbb{C}$ the map

$$\pi_\lambda : g \mapsto \Phi_g^{-\lambda} g_*$$

is a homomorphism $G \to \operatorname{End}(C^\infty(S^n))$, i.e., a representation of G. In these terms, (1.3.11) can be interpreted as the intertwining relation

$$P_{2N}(g_c) \circ \pi_{-\frac{n}{2}+N}(g) = \pi_{-\frac{n}{2}-N}(g) \circ P_{2N}(g_c), \; g \in G. \tag{1.3.12}$$

In representation theory, π_λ is known as the spherical principal series representation. More precisely, using $S^n = G/P$, π_λ can be seen as the compact realization of an *induced representation* from a parabolic subgroup P. Here, π_λ is called spherical since its restriction to the maximal compact subgroup $\mathrm{SO}(n+1)$ leaves invariant the constant function.

Now, for generic $\lambda \in \mathbb{C}$, the space $\operatorname{Hom}_G(\pi_\lambda, \pi_{-\lambda-n})$ of intertwining operators is one-dimensional. In fact, Branson [Br95] showed that

$$A_\lambda = \Gamma\left(B+\lambda+\frac{n+1}{2}\right) \Big/ \Gamma\left(B-\lambda-\frac{n-1}{2}\right) \tag{1.3.13}$$

with

$$B = \sqrt{-\Delta_{S^n} + \left(\frac{n-1}{2}\right)^2}$$

is such an operator. Here Γ denotes Euler's Gamma function. We recall that

$$\Gamma(s) = \int_0^\infty e^{-t} t^{s-1} dt$$

for $\mathrm{Re}(s) > 0$. Γ admits a meromorphic continuation to $\mathbb{C}$ with simple poles in $s = -n$, $n = 0, 1, 2, \ldots$ with residues $(-1)^n/n!$. The idea of the proof of the intertwining property of A_λ is the following. The restriction of π_λ to $K = \mathrm{SO}(n+1)$ does not depend on λ. Hence, by Schur's lemma, any intertwining operator is scalar on K-types (spherical harmonics). The analysis of the intertwining property in terms of the K-type decomposition yields the matrix coefficient

$$\Gamma(j+\lambda+n)/\Gamma(j-\lambda)$$

on the space of spherical harmonics of order j, $j \geq 0$. Now (1.3.13) follows from the fact that $-\Delta_{S^n}$ has the eigenvalues $j(n-1+j)$.

For $\lambda = -\frac{n}{2} + N$ with $N \geq 1$, (1.3.13) yields the operator

$$A_{-\frac{n}{2}+N} = (B+N-1/2)\cdots(B-N+1/2) = \prod_{j=\frac{n}{2}}^{\frac{n}{2}+N-1} (-\Delta_{S^n}+j(n-1-j)). \quad (1.3.14)$$

Thus, we find

Theorem 1.3.6. *On the round sphere S^n of dimension $n \geq 2$,*

$$P_{2N} = \prod_{j=\frac{n}{2}}^{\frac{n}{2}+N-1} (\Delta_{S^n} - j(n-1-j)) \quad (1.3.15)$$

and

$$Q_{2N} = \frac{n}{2} \prod_{j=1}^{N-1} \left(\frac{n}{2} - j\right)\left(\frac{n}{2} + j\right) \quad (1.3.16)$$

for all $N \geq 1$. In particular, $Q_n = (n-1)!$.

The following alternative formula for the intertwining operator for spherical principal series will be central in connection with the relation between GJMS-operators and scattering theory. For $\mathrm{Re}(\lambda) < -\frac{n}{2}$, we define the integral operator

$$I_\lambda(u)(x) = \int_{S^n} \frac{u(y)}{|x-y|^{2(\lambda+n)}} dy, \quad (1.3.17)$$

where $|\cdot|$ denotes the Euclidean distance and dy is the rotation invariant measure which gives S^n the volume 1.

Proposition 1.3.1. *For* $\mathrm{Re}(\lambda) < -\frac{n}{2}$, *the operator* I_λ *satisfies*

$$I_\lambda \circ \pi_\lambda(g) = \pi_{-\lambda-n}(g) \circ I_\lambda, \; g \in G. \tag{1.3.18}$$

Proof. The result is a consequence of the identity ([Ni89], equation (1.3.2))

$$|g^{-1}(x) - g^{-1}(y)|^2 = \Phi_g(x)\Phi_g(y)|x-y|^2. \tag{1.3.19}$$

In fact, the assertion is equivalent to

$$\int_{S^n} \frac{\Phi_g^{-\lambda}(y) g_*(u)(y)}{|x-y|^{2(\lambda+n)}} dy = \int_{S^n} \frac{\Phi_g^{\lambda+n}(x)u(y)}{|g^{-1}(x)-y|^{2(\lambda+n)}} dy$$

for all $u \in C^\infty(S^n)$. Under the substitution $y \mapsto g(y)$, we have $g_*(dy) = \Phi_g^n dy$, and the left-hand side transforms into

$$\int_{S^n} \frac{g^*(\Phi_g^{-\lambda-n})(y)u(y)}{|x-g(y)|^{2(\lambda+n)}} dy.$$

But (1.3.19) implies

$$|g^{-1}(x) - y|^2 = \Phi_g(x) g^*(\Phi_g)(y)|x-g(y)|^2.$$

This relation shows that the integrand equals

$$\frac{\Phi_g^{\lambda+n}(x)u(y)}{|g^{-1}(x)-y|^{2(\lambda+n)}}.$$

The proof is complete. $\square$

Note that (1.3.19) implies the well-known invariance

$$[g(x), g(y), g(z), g(w)] = [x, y, z, w]$$

of the cross-ratio

$$[x, y, z, w] = \frac{|x-z|\;|y-w|}{|x-w|\;|y-z|}.$$

Next, we relate I_λ to Branson's formula A_λ. It suffices to determine how both intertwining operators act on the constant function $u = 1$. On the one hand, we have

$$A_\lambda(1) = \frac{\Gamma(\lambda+n)}{\Gamma(-\lambda)}.$$

For $\mathrm{Re}(\lambda) < -\frac{n}{2}$, the intertwining property (1.3.18) shows that $I_\lambda(1)$ is a constant function on S^n. Thus, it is enough to calculate $I_\lambda(1)$ at $(1, 0, \ldots, 0) \in S^n$. This value is given by

$$\int_{S^n} \frac{1}{(2-2y_1)^{\lambda+n}} dy = 2^{-(\lambda+n)} \frac{\omega_{n-1}}{\omega_n} \int_0^\pi \frac{(\sin\theta)^{n-1}}{(1-\cos\theta)^{\lambda+n}} d\theta, \tag{1.3.20}$$

where ω_m denotes the volume of S^m. Now we calculate

$$\int_0^\pi \frac{(\sin\theta)^{n-1}}{(1-\cos\theta)^\lambda} d\theta = 2^{n-\lambda}\int_0^{\pi/2} (\sin\theta)^{n-1-2\lambda}(\cos\theta)^{n-1} d\theta$$
$$= 2^{n-\lambda-1} B\left(\frac{n}{2}-\lambda, \frac{n}{2}\right) = 2^{n-\lambda-1}\frac{\Gamma(\frac{n}{2}-\lambda)\Gamma(\frac{n}{2})}{\Gamma(n-\lambda)}.$$

In particular,

$$\frac{\omega_n}{\omega_{n-1}} = \int_0^\pi (\sin\theta)^{n-1} d\theta = 2^{n-1}\frac{\Gamma(\frac{n}{2})\Gamma(\frac{n}{2})}{\Gamma(n)}. \tag{1.3.21}$$

Hence $I_\lambda(1)$ equals

$$2^{-(2\lambda+n+1)}\frac{\omega_{n-1}}{\omega_n}\frac{\Gamma(\frac{n}{2})\Gamma(-\frac{n}{2}-\lambda)}{\Gamma(-\lambda)} = 2^{-(2\lambda+2n)}\frac{\Gamma(n)\Gamma(-\frac{n}{2}-\lambda)}{\Gamma(\frac{n}{2})\Gamma(-\lambda)}. \tag{1.3.22}$$

It follows that

$$\alpha(\lambda)A_\lambda = I_\lambda \tag{1.3.23}$$

if

$$\alpha(\lambda)\frac{\Gamma(\lambda+n)}{\Gamma(-\lambda)} = 2^{-(2\lambda+2n)}\frac{\Gamma(n)\Gamma(-\frac{n}{2}-\lambda)}{\Gamma(\frac{n}{2})\Gamma(-\lambda)}$$

i.e.,

$$\alpha(\lambda) = 2^{-(2\lambda+2n)}\frac{\Gamma(-\frac{n}{2}-\lambda)}{\Gamma(\lambda+n)}\frac{\Gamma(n)}{\Gamma(\frac{n}{2})}.$$

The relation (1.3.23) shows that the family I_λ admits a meromorphic continuation to $\mathbb{C}$ with simple poles at

$$\lambda \in \left\{-\frac{n}{2}+N,\ N=0,1,2,\dots\right\},$$

and that the residues of I_λ at $\lambda = -\frac{n}{2}+N$, $N \geq 1$ are constant multiples of the GJMS-operators $P_{2N}(g_c)$. More precisely, (1.3.14) and (1.3.23) yield the following result.

Proposition 1.3.2. *For any $N \geq 1$,*

$$\operatorname{Res}_{-\frac{n}{2}+N}(I_\lambda) = c_N P_{2N}(g_c) \tag{1.3.24}$$

with

$$c_N = -\frac{(-1)^N}{N!}\pi^{-\frac{1}{2}}2^{-2N-1}\frac{\Gamma(\frac{n+1}{2})}{\Gamma(\frac{n}{2}+N)}.$$

The intertwining relation (1.3.18) holds true as an identity of meromorphic families of operators. In particular, the intertwining property (1.3.12) is a consequence of Lemma 1.3.2.

Expanding I_λ into spherical harmonics and reading off residues leads to an alternative proof of the residue formula (1.3.24) (see [Be93]).

The GJMS-operators on general manifolds are much more complicated. However, for Einstein metrics the situation is similar as on the round spheres. In fact, Gover proved the following extension of Theorem 1.3.6,

Theorem 1.3.7 ([G06]). *On Einstein manifolds (M,g) of dimension $n \geq 2$,*

$$P_{2N}(g) = \prod_{j=\frac{n}{2}}^{\frac{n}{2}+N-1} \left(\Delta_g - \frac{j(n-1-j)}{n(n-1)} \tau(g) \right), \quad N \geq 1. \tag{1.3.25}$$

In particular, for even n, the critical Q-curvature is given by

$$Q_n(g) = (n-1)! \left(\frac{\tau(g)}{n(n-1)} \right)^{\frac{n}{2}}. \tag{1.3.26}$$

The intertwining operator I_λ on $C^\infty(S^n)$ can be regarded as the geometric scattering operator of the Poincaré-metric on the unit ball $\mathbb{B}^{n+1}$ with the (conformal) boundary S^n. This extrinsic perspective towards I_λ resembles the extrinsic type of construction of GJMS-operators. In fact, in Section 1.4 we shall describe the profound relations between both.

1.4 Scattering theory

In the present section, we describe GJMS-operators and critical Q-curvature of a manifold of even dimension in terms of geometric scattering theory [Me95] of the Laplacian of the Poincaré-metric g_+ (an Einstein metric with negative cosmological constant) on a space of one more dimension.

The results extend the relation between GJMS-operators of the round sphere and the scattering operator of the Laplacian of the ball model of hyperbolic space.

We start with a description of I_λ on the sphere as the scattering operator of the hyperbolic unit ball. For this purpose, we consider the ball

$$\mathbb{B}^{n+1} = \{ x \in \mathbb{R}^{n+1} \mid |x| < 1 \}$$

with the metric

$$g_h = \frac{4}{(1-|x|^2)^2} (dx_1^2 + \cdots + dx_{n+1}^2).$$

Moreover, we consider eigenfunctions $u \in C^\infty(\mathbb{B}^{n+1})$ of the Laplacian Δ_h:

$$-\Delta_h u = \lambda(n-\lambda)u. \tag{1.4.1}$$

By the invariance of the Laplacian with respect to isometries, the space of solutions of (1.4.1) is an $\mathrm{SO}(1,n)^\circ$-module. In fact, for generic λ, this module is isomorphic to the hyperfunction globalization of a spherical principal series [Sch85]. Moreover, the isomorphisms are provided by Helgason's Poisson transforms [H08]. This means, that, for generic λ, any eigenfunction u is the Poisson-transform $\mathcal{P}_\lambda(\omega)$ of a hyperfunction $\omega \in \mathcal{A}'(S^n)$, and the Poisson-transforms $\mathcal{P}_\lambda$ intertwine push-forward on functions on the ball with spherical principal series representations. It is natural to think of ω as a "boundary value" of u. Thus, there is an equivariant way to associate "boundary values" to eigenfunctions. In fact, for any generic λ, there are actually *two* such ways by using either $\mathcal{P}_\lambda$ or $\mathcal{P}_{n-\lambda}$. In other words, any eigenfunction has *two* natural boundary values. The linear operator which relates these boundary values is the *geometric scattering operator*. By the equivariance of the Poisson-transforms, the scattering operator is an intertwining operator for spherical principal series, i.e., a multiple of I_λ.

Now we make the above picture more explicit. Helgason's Poisson transform $\mathcal{P}_\lambda$ is defined by the kernel

$$P(x,y)^\lambda = \left(\frac{1-|x|^2}{|x-y|^2}\right)^\lambda, \quad x \in \mathbb{B}^{n+1}, y \in S^n.$$

Then, for generic λ, the solutions u of (1.4.1) admit a representation

$$u(x) = \mathcal{P}_\lambda(\omega)(x) = \int_{S^n} P(x,y)^\lambda \omega(y)dy. \tag{1.4.2}$$

For hyperfunctions (or distributions) ω, the integral, of course, is to be understood as the pairing

$$\langle P(x,\cdot)^\lambda, \omega(\cdot)\rangle.$$

It is well-defined for real-analytic ω since the kernel P^λ is real-analytic in the boundary variable y. For $n = 1$ and $\lambda = 1$, the representation formula (1.4.2) specializes to the familiar representation

$$u(x) = \int_{S^1} \frac{1-|x|^2}{|x-y|^2}\omega(y)dy$$

of harmonic functions for the Euclidean metric; recall that in dimension $n = 2$ the conformal covariance of the Laplacian implies that the two notions of harmonic functions with respect to the hyperbolic and the Euclidean metric coincide.

The Poisson-transforms map (generalized) functions on the sphere to eigenfunctions of the Laplacian on the ball. The result that *all* eigenfunctions are Poisson-transforms is a consequence of the existence of asymptotic developments of eigenfunctions near the boundary such that the boundary values can be read off from the leading terms of the developments.

In the following, we shall restrict attention to eigenfunctions with *smooth* boundary values. In terms of polar coordinates $x = ry$ with $0 < r < 1$ and $y \in S^n$, these have asymptotic expansions of the form

$$u(ry) \sim \sum_{N\geq 0} (1-r^2)^{\lambda+N} a_N(\lambda,y) + \sum_{N\geq 0} (1-r^2)^{n-\lambda+N} b_N(\lambda,y), \; r \to 1. \tag{1.4.3}$$

The coefficients in the expansion (1.4.3) form two sequences

$$a_0, a_1, a_2, \ldots \quad \text{and} \quad b_0, b_1, b_2, \ldots$$

of smooth function on S^n. The basic structural properties of (1.4.3) are the following.

- u is determined by one of the leading coefficients a_0 or b_0. These are (multiples) of the boundary values of u.
- $a_0 = \mathcal{S}(\lambda)(b_0)$, where $\mathcal{S}(\lambda)$ is the geometric scattering operator.
- The coefficients a_N, b_N are given by two sequences of rational families of differential operators acting on the respective terms a_0, b_0.
- Although the individual coefficients $a_N(\lambda;\cdot)$, $b_N(\lambda,\cdot)$ are only rational, the poles cancel in the sum.

We shall not go into the formalities of the theory. These can be found in [H08]. Instead, we shall illustrate these features in terms of sufficiently suggestive special cases. We consider the rotation-invariant eigenfunction u on $\mathbb{B}^2$ which is given by

$$u(r) = \mathcal{P}_\lambda(1)(r) = (1-r^2)^\lambda \frac{1}{2\pi} \int_0^{2\pi} (r^2 - 2r\cos\theta + 1)^{-\lambda} d\theta, \; r < 1.$$

u can be written in terms of Gauß's hypergeometric functions [BE53] as

$$u(r) = (1-r^2)^\lambda F(\lambda, \lambda; 1; r^2), \tag{1.4.4}$$

where

$$F(a,b;c;x) = \sum_{n \geq 0} \frac{(a)_n (b)_n}{(c)_n} \frac{x^n}{n!}, \; |x| < 1.$$

For the proof of (1.4.4), we write

$$(r^2 - 2r\cos\theta + 1)^{-\lambda} = (1 - re^{i\theta})^{-\lambda} 1 - re^{-i\theta})^{-\lambda}$$

and decompose the right-hand side as

$$\left(\sum_{j \geq 0} \frac{(\lambda)_j}{j!} r^j e^{ij\theta} \right) \left(\sum_{k \geq 0} \frac{(\lambda)_k}{k!} r^k e^{ik\theta} \right).$$

Then (1.4.4) follows by direct calculation. In order to find the asymptotic expansion of u for $r \to 1$, we apply the identity ([BE53], 2.10, (1))

$$\begin{aligned} F(a,b;c;r) = {} & AF(a,b;a+b-c+1;1-r) \\ & + B(1-r)^{c-a-b} F(c-a, c-b; c-a-b+1; 1-r) \end{aligned} \tag{1.4.5}$$

with

$$A = \frac{\Gamma(c)\Gamma(c-a-b)}{\Gamma(c-a)\Gamma(c-b)} \quad \text{and} \quad B = \frac{\Gamma(c)\Gamma(a+b-c)}{\Gamma(a)\Gamma(b)}.$$

We find

$$F(\lambda,\lambda;1;r) = A(\lambda)F(\lambda,\lambda;2\lambda;1-r) \\ + B(\lambda)(1-r^2)^{1-2\lambda}F(1-\lambda,1-\lambda;2-2\lambda;1-r)$$

with

$$A(\lambda) = B(1-\lambda) \quad \text{and} \quad B(\lambda) = \frac{\Gamma(2\lambda-1)}{\Gamma(\lambda)\Gamma(\lambda)} = \pi^{-\frac{1}{2}}2^{2\lambda-2}\frac{\Gamma(\lambda-\frac{1}{2})}{\Gamma(\lambda)}.$$

Now the power series expansion of F yields the desired expansion of u. For the leading terms we find

$$a_0(\lambda) = A(\lambda) = B(1-\lambda) \quad \text{and} \quad b_0(\lambda) = B(\lambda).$$

Using (1.3.22), we write this result in the form

$$a_0(\lambda) = I_{\lambda-1}(1) \quad \text{and} \quad b_0(\lambda) = c(\lambda), \tag{1.4.6}$$

where

$$c(\lambda) = \pi^{-\frac{1}{2}}2^{2\lambda-2}\frac{\Gamma(\lambda-\frac{1}{2})}{\Gamma(\lambda)}.$$

More generally, for any $\omega \in C^\infty(S^1)$, one of the leading terms of the asymptotic expansion of the eigenfunction $\mathcal{P}_\lambda(\omega) \in C^\infty(\mathbb{B}^2)$ is of the form

$$b_0(\lambda) = c(\lambda)\omega.$$

Now the Poisson transform $\mathcal{P}_\lambda$ satisfies the intertwining relations

$$g_* \circ \mathcal{P}_\lambda = \mathcal{P}_\lambda \circ \pi_{\lambda-1}(g). \tag{1.4.7}$$

(1.4.7) is a consequence of the identity

$$P(g\cdot 0, y) = g_*(\mathrm{vol}(g_c))/\mathrm{vol}(g_c). \tag{1.4.8}$$

In order to prove (1.4.8), we first note that, for $g = \begin{pmatrix} d & c \\ b^t & A \end{pmatrix}$, (1.1.7) implies

$$g_*(\mathrm{vol}(g_c))/\mathrm{vol}(g_c) = 1/(d-(b,y)),\ y \in S^1.$$

On the other hand, $g\cdot 0 = b/(1+d)$ by (1.1.9). Hence

$$P(g\cdot 0, y) = \frac{1-|g\cdot 0|^2}{|g\cdot 0 - y|^2} = \frac{(1+d)^2-|b|^2}{|b-(1+d)y|^2},$$

and an easy calculation using $d^2 - |b|^2 = 1$ proves the assertion.

(1.4.7) implies that the composition

$$\mathcal{P}_{\lambda-1}^{-1} \circ \mathcal{P}_\lambda : C^\infty(S^1) \to C^\infty(S^1)$$

is an intertwining operator in $\mathrm{Hom}_G(\pi_{\lambda-1}, \pi_{-\lambda})$. Hence the other leading term in the asymptotic expansion of $\mathcal{P}_\lambda(\omega)$ is a multiple

$$\kappa(\lambda) I_{\lambda-1}(\omega)$$

of $I_{\lambda-1}(\omega)$. In particular, for $\omega = 1$, it yields $\kappa(\lambda) I_{\lambda-1}(1)$. Hence the first relation in (1.4.6) implies $\kappa(\lambda) = 1$.

Now we define the geometric scattering operator $\mathcal{S}(\lambda)$ by the relation

$$a_0 = \mathcal{S}(\lambda)(b_0) \tag{1.4.9}$$

of the leading terms in the asymptotic expansions of eigenfunctions of the Laplacian. The definition immediately implies that the scattering operator $\mathcal{S}(\lambda)$ satisfied the functional equation

$$\mathcal{S}(\lambda) \circ \mathcal{S}(1-\lambda) = 1. \tag{1.4.10}$$

Using $\mathcal{S}(\lambda)(c(\lambda)) = I_{\lambda-1}(1)$, we obtain

$$\mathcal{S}(\lambda) = \frac{I_{\lambda-1}}{c(\lambda)}. \tag{1.4.11}$$

This relation shows that $\mathcal{S}(\lambda)$ has simple poles in $\lambda = \frac{1}{2} + N$, $N \geq 1$.

Note that in terms of I_λ and $c(\lambda)$, the functional equation for $\mathcal{S}(\lambda)$ reads

$$I_{\lambda-1} \circ I_{-\lambda} = c(\lambda) c(1-\lambda).$$

Note also that, by (1.3.23), the relation (1.4.11) is equivalent to

$$\mathcal{S}(\lambda) = 2^{2(1-2\lambda)} \frac{\Gamma(\frac{1}{2}-\lambda)}{\Gamma(\lambda-\frac{1}{2})} A_{\lambda-1}.$$

The above results extend as follows to dimension $n \geq 2$. The $\mathrm{SO}(n+1)$-invariant eigenfunction $u = \mathcal{P}_\lambda(1) \in C^\infty(\mathbb{B}^{n+1})$ is determined by $u(r) = \mathcal{P}_\lambda(1)(r)$, $r < 1$. But

$$\begin{aligned} u(r) &= (1-r^2)^\lambda \int_{S^n} |(r, 0, \ldots, 0) - y|^{-2\lambda} dy \\ &= (1-r^2)^\lambda \left(\frac{\omega_{n-1}}{\omega_n}\right) \int_0^\pi (1 - 2r\cos\theta + r^2)^{-\lambda} (\sin\theta)^{n-1} d\theta \\ &= (1-r^2)^\lambda \left(\frac{\omega_{n-1}}{\omega_n}\right) \frac{\Gamma(\frac{1}{2})\Gamma(\frac{n}{2})}{\Gamma(\frac{n+1}{2})} F\left(\lambda, \lambda - \frac{n-1}{2}; \frac{n+1}{2}; r^2\right) \end{aligned}$$

by [BE53], 2.4, (9). Using (1.3.21) and Legendre's duplication formula, we find

$$u(r) = (1-r^2)^\lambda F\left(\lambda, \lambda - \frac{n-1}{2}; \frac{n+1}{2}; r^2\right). \tag{1.4.12}$$

Now (1.4.5) shows that the leading terms of the expansion of $u(r)$ into powers of $(1-r^2)$ are given by

$$a_0(\lambda) = \frac{\Gamma(\frac{n+1}{2})\Gamma(n-2\lambda)}{\Gamma(\frac{n+1}{2}-\lambda)\Gamma(n-\lambda)} \quad \text{and} \quad b_0(\lambda) = \frac{\Gamma(\frac{n+1}{2})\Gamma(2\lambda-n)}{\Gamma(\lambda)\Gamma(\lambda-\frac{n-1}{2})}.$$

Note that by Legendre's duplication formula

$$b_0(\lambda) = c(\lambda),$$

where

$$c(\lambda) = \pi^{-\frac{1}{2}} 2^{2\lambda-n-1} \frac{\Gamma(\frac{n+1}{2})\Gamma(\lambda-\frac{n}{2})}{\Gamma(\lambda)} \tag{1.4.13}$$

is known as Harish-Chandra's c-function. Thus, the expansion of u has the form

$$u(r) = \left(I_\lambda(1)(1-r^2)^\lambda + \cdots\right) + \left(c(\lambda)(1-r^2)^{n-\lambda} + \cdots\right),$$

and the scattering operator is given by

$$\mathcal{S}(\lambda) = \frac{I_{\lambda-n}}{c(\lambda)}. \tag{1.4.14}$$

Now Proposition 1.3.2 implies that $\mathcal{S}(\lambda)$ is a meromorphic family with simple poles in $\lambda = \frac{n}{2} + N$, $N = 1, 2, \ldots$. A calculation yields the following explicit formula for the residues.

Proposition 1.4.1. *The residues of $\mathcal{S}(\lambda)$ in $\frac{n}{2} + N$, $N = 1, 2, \ldots$ are given by the formula*

$$\operatorname{Res}_{\frac{n}{2}+N}(\mathcal{S}(\lambda)) = -\frac{1}{N!(N-1)!2^{2N}} P_{2N}(4g_c). \tag{1.4.15}$$

It is natural to express the right-hand side of (1.4.15) in terms of the operators P_{2N} for the metric $4g_c$ since the conformal compactification

$$(1-|x|^2)^2 g_h$$

of the hyperbolic metric g_h pulls back to $4g_c$ under $S^n \hookrightarrow \mathbb{R}^{n+1}$. This will become clearer below.

Next, we analyze the structure of the coefficients a_N and b_N for $N \geq 1$. The main observation is that the sum

$$\sum_{N\geq 0} (1-r^2)^{\lambda+N} a_N(\lambda)$$

is a formal solution of

$$-\Delta_h u = \lambda(n-\lambda)u$$

if and only if the coefficients satisfy a recursive relation for the coefficients. In fact, a routine calculation shows that

$$\begin{aligned}[(k+1)(n-1-2\lambda-k)]a_{k+1} \\ = \frac{1}{2}[(\lambda+k)(n-1-2(\lambda+k))]a_k + \frac{1}{4}\Delta_{S^n}(a_{k-1}+\cdots+a_0),\ k\geq 0. \quad (1.4.16)\end{aligned}$$

In particular, we find

$$a_1 = \frac{1}{2}\lambda a_0$$

and

$$a_2 = \frac{1}{8(n-2-2\lambda)}\left(\Delta_{S^n} + \lambda(\lambda+1)(n-3-2\lambda)\right)a_0 = \mathcal{T}_2(\lambda)(a_0). \quad (1.4.17)$$

We consider $\mathcal{T}_2(\lambda)$ as a meromorphic family of operators. (1.4.17) shows that $\mathcal{T}_2(\lambda)$ has a simple pole in $\lambda = \frac{n}{2}-1$. More generally, the analogous operator $\mathcal{T}_{2N}(\lambda)$ which yields $a_{2N}(\lambda)$ in terms of a_0 has a simple pole in $\lambda = \frac{n}{2}-N$.

Now we observe that

$$\operatorname{Res}_{\frac{n}{2}-1}(\mathcal{T}_2(\lambda)) = -\frac{1}{16}\left(\Delta_{S^n} - \left(\frac{n}{2}-1\right)\frac{n}{2}\right) = -\frac{1}{4}P_2(4g_c), \quad (1.4.18)$$

i.e., the residue of $\mathcal{T}_2(\lambda)$ at $\lambda = \frac{n}{2}-1$ yields the Yamabe operator of $(S^n, 4g_c)$. The *conceptual* explanation of this observation is as follows. We consider the asymptotic expansion

$$u(ry) \sim \sum_{j\geq 0}(1-r^2)^{\lambda+j}a_j(\lambda,y) + \sum_{j\geq 0}(1-r^2)^{n-\lambda+j}b_j(\lambda,y),\ y\in S^n$$

of an eigenfunction $u = \mathcal{P}_\lambda(f)$ with $f \in C^\infty(S^n)$ (see (1.4.3)). The coefficients $a_j(\lambda,\cdot) \in C^\infty(S^n)$ and $b_j(\lambda,\cdot) \in C^\infty(S^n)$ in the expansion are rational in λ. On the other hand, u is holomorphic in λ. Thus, the residues of the expansion vanish. In particular, for an eigenfunction with boundary value b_0, we have $a_0 = \mathcal{S}(\lambda)(b_0)$ and the residue

$$\operatorname{Res}_{\frac{n}{2}+1}(\mathcal{S}(\lambda)(b_0)) + \operatorname{Res}_{\frac{n}{2}+1}(\mathcal{T}_2(n-\lambda)(b_0)) \quad (1.4.19)$$

vanishes. Now Proposition 1.4.1 implies the formula

$$\operatorname{Res}_{\frac{n}{2}+1}(\mathcal{T}_2(n-\lambda)) = \frac{1}{4}P_2(4g_c),$$

i.e.,

$$\operatorname{Res}_{\frac{n}{2}-1}(\mathcal{T}_2(\lambda)) = -\frac{1}{4}P_2(4g_c).$$

Similarly, Proposition 1.4.1 yields the relations

$$\operatorname{Res}_{\frac{n}{2}-N}(\mathcal{T}_{2N}(\lambda)) = -\frac{1}{N!(N-1)!2^{2N}} P_{2N}(4g_c) \tag{1.4.20}$$

for all $N \geq 1$.

As noted above, the metric $4g_c$ on the right-hand side of (1.4.20) is to be interpreted as the pull-back to S^n of the conformal compactification

$$(1-|x|^2)^2 g_h \tag{1.4.21}$$

of the hyperbolic metric g_h. Other conformal compactifications lead to analogs of (1.4.20). Next, we describe one of these which later will be of particular importance in a much broader context. The metric

$$\bar{g}_h = \left(\frac{1-r}{1+r}\right)^2 g_h \tag{1.4.22}$$

is smooth up to the boundary S^n and

$$i^*(\bar{g}_h) = \frac{1}{4} g_{S^n}.$$

We expand eigenfunctions in the form

$$u(ry) \sim \sum_{N\geq 0} \left(\frac{1-r}{1+r}\right)^{\lambda+N} a_N(\lambda,y) + \sum_{N\geq 0} \left(\frac{1-r}{1+r}\right)^{n-\lambda+N} b_N(\lambda,y)$$

for $y \in S^n$, $r \to 1$. In view of

$$\frac{1-|x|}{1+|x|} = e^{-d(x,0)},$$

where d denotes the hyperbolic distance, we can write the expansion also in the form

$$u((\tanh r)y) \sim \sum_{N\geq 0} e^{-2r(\lambda+N)} a_N(\lambda,y) + \sum_{N\geq 0} e^{-2r(n-\lambda+N)} b_N(\lambda,y), \quad r\to\infty.$$

This form of radial asymptotic expansions is naturally adapted to the Cartan decomposition $G = K\overline{A^+}K$ [H08] and therefore widely used in harmonic analysis. Now the substitution

$$s = \frac{1-r}{1+r}$$

yields

$$g_h = \frac{1}{s^2}\left(ds^2 + \left(\frac{1-s^2}{2}\right)^2 g_{S^n}\right)$$

and the conformal compactification (1.4.22) reads

$$\bar{g}_h = s^2 g_h = ds^2 + \left(\frac{1-s^2}{2}\right)^2 g_{S^n} = ds^2 + \frac{1}{4} g_{S^n} - \frac{s^2}{2} g_{S^n} + \frac{s^4}{4} g_{S^n}. \tag{1.4.23}$$

The formal expansion

$$\sum_{N\geq 0} s^{\lambda+N} \bar{a}_N(\lambda, y) + \sum_{N\geq 0} s^{n-\lambda+N} \bar{b}_N(\lambda, y), \; s \to 0 \tag{1.4.24}$$

of an eigenfunction has the property that the coefficients $\bar{a}_{\text{odd}}$ and $\bar{b}_{\text{odd}}$ vanish. Moreover, in a formal calculation, the leading coefficients $\bar{a}_0$ and $\bar{b}_0$ are free, and all higher coefficients are determined by certain rational families of linear differential operators acting on the leading coefficients. Moreover, for genuine eigenfunctions, $\bar{a}_0$ and $\bar{b}_0$ are related by a scattering operator:

$$\bar{a}_0 = \bar{\mathcal{S}}(\lambda)(\bar{b}_0). \tag{1.4.25}$$

Now explicit calculations yield

$$\bar{a}_2(\lambda) = \frac{1}{2(n-2-2\lambda)}(4\Delta_{S^n} - 2\lambda n)(a_0) = \overline{\mathcal{T}}_2(\lambda)(\bar{a}_0) \tag{1.4.26}$$

(and a similar formula for $\bar{b}_2(\lambda)$). The latter formula shows that

$$\operatorname{Res}_{\frac{n}{2}-1}(\overline{\mathcal{T}}_2(\lambda)) = -\left(\Delta_{S^n} - \frac{n}{2}\left(\frac{n}{2}-1\right)\right) = -P_2(S^n, g_c)$$

i.e.,

$$\operatorname{Res}_{\frac{n}{2}-1}(\overline{\mathcal{T}}_2(\lambda)) = -\frac{1}{4} P_2(S^n, g_c/4).$$

This relation is the analog of (1.4.18).

The scattering operators $\mathcal{S}(\lambda)$ and $\bar{\mathcal{S}}(\lambda)$, which relate the respective leading terms in the expansions (1.4.3) and (1.4.24) of an eigenfunction, are conjugate:

$$\bar{\mathcal{S}}(\lambda) = 2^{2\lambda} \circ \mathcal{S}(\lambda) \circ 2^{-2(n-\lambda)}. \tag{1.4.27}$$

It is natural to regard $\mathcal{S}(\lambda)$ and $\bar{\mathcal{S}}(\lambda)$ as operators which are associated to the respective metrics

$$i^*((1-r^2)^2 g_h) = 4g_c = h \quad \text{and} \quad i^*\left(\left(\frac{1-r}{1+r}\right)^2 g_h\right) = \frac{1}{4} g_c = \bar{h}$$

on S^n. These are conformally equivalent: $\bar{h} = e^{2\varphi} h$ with $e^{\varphi} = 2^{-2}$. In these terms, (1.4.27) can be written in the form

$$\mathcal{S}(h; \lambda) = e^{-\lambda\varphi} \circ \mathcal{S}(\bar{h}; \lambda) \circ e^{(n-\lambda)\varphi}.$$

Note that the consequence

$$\operatorname{Res}_{\frac{n}{2}+1}(\bar{\mathcal{S}}) = 2^{n+2} \circ \operatorname{Res}_{\frac{n}{2}+1}(\mathcal{S}) \circ 2^{-n+2}$$

is equivalent to the relation $P_2(\bar{h}) = 2^{n+2} \circ P_2(h) \circ 2^{-n+2}$ which is a special case of the conformal covariance of P_2. We omit the details concerning the analogous observations for the residues of the scattering operators at $\lambda = \frac{n}{2} + N$ for $N \geq 2$.

The discussion shows that the residues of scattering operators yield GJMS-operators P_{2N} for certain metrics in the conformal class of the round metric. Moreover, we have seen that different choices of the asymptotic expansions are reflected in the conformal covariance of the residues.

After these preparations, we continue with the description of results of Graham and Zworski [GZ03] which, roughly speaking, state that the relation between GJMS-operators and residues of scattering operators on round spheres extends to *all* Riemannian manifolds.

We start by defining the setting. The pair

$$(\mathbb{B}^{n+1}, S^n)$$

with the hyperbolic metric g_h on the open ball $\mathbb{B}^{n+1}$ and the conformal class $[g_c]$ of the round metric on S^n will be replaced by the pair

$$(X^{n+1}, M^n)$$

consisting of the open interior X of a compact manifold $\bar{X}$ with boundary M with a Poincaré-Einstein metric g_+ on X and an induced conformal class $[h]$ on M.

Definition 1.4.1. A metric g on X is called conformally compact if there exists a defining function $\rho \in C^\infty(\bar{X})$ of the boundary M, i.e., $\rho > 0$ on X, $\rho = 0$ on M and $d\rho \neq 0$ on M, so that $\rho^2 g$ extends to a smooth metric up to the boundary.

In the situation of Definition 1.4.1, the metrics $\bar{g} = \rho^2 g$ are called *conformal compactifications* of g. Any conformal compactification induces a metric on M by

$$i^*(\rho^2 g),$$

where $i : M \hookrightarrow \bar{X}$. A change of the defining function yields a metric in the same conformal class on the boundary. The conformal class

$$c = [i^*(\rho^2 g)]$$

is called the *conformal infinity* of g. The function $|d\rho|^2_{\bar{g}}$ on M does not depend on the choice of ρ. In fact, the value of $|d\rho|^2_{\bar{g}}$ at a point of M is the negative of the asymptotic sectional curvature of g at this point. In particular, g is asymptotically hyperbolic, i.e., the sectional curvatures approach -1, iff

$$|d\rho|_{\bar{g}} = 1 \text{ on } M.$$

Conversely, for a given asymptotically hyperbolic g with conformal infinity c and any choice of a metric $h \in c$, there exists a *unique* defining function $\rho \in C^\infty(X)$ so that

$$|d\rho|^2_{\rho^2 g} = 1 \text{ near } M \tag{1.4.28}$$

and

$$i^*(\rho^2 g) = h.$$

Here the idea is that (1.4.28) is valid not only on the boundary but in a neighbourhood. Now using the gradient flow of ρ (with respect to the metric $\rho^2 g$) we find coordinates so that g takes the normal form

$$g = \rho^{-2}(d\rho^2 + h_\rho) \tag{1.4.29}$$

for a one-parameter family h_ρ of metrics with $h_0 = h$.

In the following, we will be interested in *special* asymptotically hyperbolic metrics: Poincaré-Einstein metrics. One of the main features of these metrics on X is that, near the boundary M, they are determined by their conformal infinities (at least to some extent). For more details concerning the following discussion we refer to [FG07].

Definition 1.4.2. A Poincaré-Einstein metric, or just Poincaré metric, for (M, c) is a conformally compact metric g_+ on X so that

1. g_+ has conformal infinity c.
2. For odd n, $\mathrm{Ric}(g_+) + ng_+$ vanishes to infinite order along M.
3. For even $n \geq 4$,
$$\mathrm{Ric}(g_+) + ng_+ = O(\rho^{n-2}) \tag{1.4.30}$$
and
$$\mathrm{tr}_h(i^*(\rho^{-(n-2)}(\mathrm{Ric}(g_+) + ng_+))) = 0, \; h \in c. \tag{1.4.31}$$

Note that the hyperbolic metric on $\mathbb{B}^{n+1}$ is a Poincaré-Einstein metric with conformal infinity given by the conformal class of the round metric.

The conditions in Definition 1.4.2 suffice to prove existence and uniqueness (up to diffeomorphisms which fix the boundary) for *even* Poincaré metrics on $X = M \times (0, \varepsilon)$. Full details are given in Chapter 4 of [FG07]. Such results rest on the local analysis of the condition (1.4.30). In fact, an analysis of this condition for a metric of the form

$$g_+ = r^{-2}(dr^2 + h_r),$$

which is given by a one-parameter family h_r of metrics such that $h_0 = h$, leads to the following structural results.

The quality of the situation depends on the parity of the dimension n. For odd n,

$$h_r = h + \underbrace{r^2 h_{(2)} + \cdots + r^{n-1} h_{(n-1)}}_{\text{even powers}} + r^n h_{(n)} + \cdots . \tag{1.4.32}$$

The Taylor coefficients $h_2, \dots, h_{(n-1)}$ are determined inductively, and are given by polynomial formulas in terms of h, its inverse, the curvature tensor of h and its covariant derivatives. The term $h_{(n)}$ is trace-free but otherwise undetermined. If h_r is assumed to be even, then $h_{(n)} = 0$ and all higher terms are determined.

Similarly, for even n,

$$h_r = h + \underbrace{r^2 h_{(2)} + \cdots + r^{n-2} h_{(n-2)}}_{\text{even powers}} + r^n (h_{(n)} + \log r \bar{h}_{(n)}) + \cdots . \tag{1.4.33}$$

The Taylor coefficients $h_2, \dots, h_{(n-2)}$ are determined inductively, and are given by polynomial formulas in terms of h, its inverse, the curvature tensor of h and its covariant derivatives. The term $\bar{h}_{(n)}$ and the quantity $\operatorname{tr} h_{(n)}$ are determined. Moreover, $\bar{h}_{(n)}$ is trace-free, and the trace-free part of $h_{(n)}$ is undetermined.

The following result describes the first two terms in these expansions.

Theorem 1.4.1. *Let $n \geq 3$. Then, the first two terms in the Taylor expansion of the Poincaré-Einstein metric $r^{-2}(dr^2 + h_r)$ with $h_0 = h$ are*

$$h_{(2)} = -\mathsf{P},$$
$$h_{(4)} = \frac{1}{4(n-4)}((n-4)\mathsf{P}^2 - \mathcal{B}).$$

Here P is the Schouten tensor of h, and

$$\mathcal{B}_{ij} = \Delta(\mathsf{P})_{ij} - \nabla^k \nabla_j (\mathsf{P})_{ik} + \mathsf{P}^{kl} \mathsf{W}_{kijl} \tag{1.4.34}$$

generalizes the Bach tensor (1.2.20) *to general dimensions.*

Here we regard P also as an endomorphisms of TM, and identify its square P^2 with a symmetric bilinear form.

Now, spectral theory of the Laplacian of a Poincaré-Einstein metric g_+ on X naturally gives rise to a scattering operator

$$\mathcal{S}(h; \lambda) : C^\infty(M) \to C^\infty(M).$$

We recall that the spectrum $\sigma(-\Delta_g)$ of the Laplacian of an asymptotically hyperbolic metric g is the union of a *finite* set of L^2-eigenvalues contained in $(0, (\frac{n}{2})^2)$ and an absolutely continuous spectrum $[(\frac{n}{2})^2, \infty)$ of infinite multiplicity [MM87]. The continuous spectrum gives rise to a scattering operator $\mathcal{S}(\lambda)$ as follows. Let $\operatorname{Re}(\lambda) = \frac{n}{2}$ and $\lambda \neq \frac{n}{2}$. We consider solutions (generalized eigenfunctions) of the equation

$$-\Delta_g u = \lambda(n-\lambda) u$$

which are of the form

$$u = F \rho^{n-\lambda} + G \rho^\lambda \quad \text{with} \quad F, G \in C^\infty(\bar{X}). \tag{1.4.35}$$

The restrictions of F and G to M are regarded as incoming and outgoing scattering data, respectively, and the scattering operator is defined as

$$\mathcal{S}(\lambda) : F|_M \mapsto G|_M.$$

This yields a well-defined operator on $C^\infty(M)$ since for any given $f \in C^\infty(M)$ there exists a unique generalized eigenfunction u with (1.4.35) and $F|_M = f$. In fact, for $\mathrm{Re}(\lambda) = \frac{n}{2}$, $\lambda \neq \frac{n}{2}$, Graham and Zworski [GZ03] construct a family

$$\mathcal{P}(\lambda) : C^\infty(M) \to \ker(\Delta_g + \lambda(n-\lambda))$$

of operators with the desired properties. $\mathcal{P}(\lambda)$ has a meromorphic continuation to $\mathrm{Re}(\lambda) > \frac{n}{2}$ with poles only for those λ such that $\lambda(n-\lambda) \in \sigma_d$.

The family $\mathcal{P}(\lambda)$ is a curved analog of Helgason's Poisson transform.

Applying complex conjugation to (1.4.35) shows the functional equation

$$\mathcal{S}(\lambda) \circ \mathcal{S}(n-\lambda) = 1$$

for $\mathrm{Re}(\lambda) = \frac{n}{2}$ so that $\lambda \neq \frac{n}{2}$. Moreover, $\mathcal{S}(\lambda)$ is unitary and regular on the line $\mathrm{Re}(\lambda) = \frac{n}{2}$, and Schwarz reflection yields a meromorphic continuation to $\mathbb{C}$ which satisfies

$$\mathcal{S}(n-\bar{\lambda})^* = \mathcal{S}(\lambda)^{-1} \quad \text{and} \quad \mathcal{S}(\lambda) \circ \mathcal{S}(n-\lambda) = 1.$$

In particular, for real λ,

$$\mathcal{S}(\lambda)^* = \mathcal{S}(n-\lambda)^{-1} = \mathcal{S}(\lambda). \tag{1.4.36}$$

Obviously, $\mathcal{S}(\lambda)$ depends on the choice of the defining function ρ. However, in the setting of a Poincaré-Einstein metric with conformal infinity c, the choice of a metric $h \in c$, uniquely determines a defining function ρ, and we denote the resulting scattering operator by $\mathcal{S}(h;\lambda)$. The following conformal transformation law is a direct consequence of the definitions.

Proposition 1.4.2. *The scattering operator $\mathcal{S}(\cdot;\lambda)$ is conformally covariant in the sense that*

$$\mathcal{S}(e^{2\varphi}h;\lambda) = e^{-\lambda\varphi} \circ \mathcal{S}(h;\lambda) \circ e^{(n-\lambda)\varphi}$$

for all $\varphi \in C^\infty(M)$.

The following theorem is the main result of [GZ03].

Theorem 1.4.2. *The scattering operator $\mathcal{S}(h;\lambda)$ is meromorphic in* $\mathrm{Re}(\lambda) > \frac{n}{2}$. *Assume that for $\lambda = \frac{n}{2} + N$ with*

$$N \in \begin{cases} 1, 2, \ldots, \frac{n}{2} & n \text{ even} \\ \mathbb{N} & \text{else} \end{cases}$$

the value $\lambda(n-\lambda)$ does not belong to the discrete spectrum σ_d of $-\Delta_g$. Then $\mathcal{S}(h;\lambda)$ has a simple pole at $\lambda = \frac{n}{2} + N$ and

$$\operatorname{Res}_{\frac{n}{2}+N}(\mathcal{S}(h;\lambda)) = -c_N P_{2N}(h) \tag{1.4.37}$$

with

$$c_N = \frac{1}{2^{2N} N!(N-1)!}.$$

Note that the conformal covariance of P_{2N} is a consequence of the conformal transformation law of $\mathcal{S}$ (Proposition 1.4.2).

Corollary 1.4.1. *The GJMS-operators P_{2N} are formally selfadjoint.*

Proof. Combine (1.4.37) with (1.4.36). □

For even n, $\mathcal{S}(h;\lambda)$ has a simple pole at $\lambda = n$. Thus, for λ near n, $\mathcal{S}(h;\lambda)$ has the form

$$\mathcal{S}(h;\lambda) = -c_{\frac{n}{2}} \frac{P_n(h)}{\lambda - n} + \mathcal{S}_0(h;\lambda) \tag{1.4.38}$$

with a holomorphic family $\mathcal{S}_0(h;\lambda)$. But since $P_n(h)$ annihilates constants, we find

$$\operatorname{Res}_n(\mathcal{S}(h;\lambda)(1)) = -c_{\frac{n}{2}} P_n(h)(1) = 0,$$

i.e., the function $\lambda \mapsto \mathcal{S}(h;\lambda)(1)$ is regular at $\lambda = n$. In particular, the quantity $\mathcal{S}(h;n)(1) \in C^\infty(M)$ is well-defined. Now the conformal transformation law of $\mathcal{S}$ (Proposition 1.4.2) implies

$$\mathcal{S}(h;\lambda)(e^{(n-\lambda)\varphi}) = e^{\lambda\varphi} \mathcal{S}(e^{2\varphi}h;\lambda)(1).$$

We combine this relation with (1.4.38) and find

$$c_{\frac{n}{2}} \lim_{\lambda \to n} P_n(h) \left(\frac{e^{(n-\lambda)\varphi}}{n-\lambda} \right) + \mathcal{S}_0(h;n)(1) = e^{n\varphi} \mathcal{S}_0(e^{2\varphi}h;n)(1).$$

Hence

$$c_{\frac{n}{2}} P_n(h)(\varphi) + \mathcal{S}_0(h;n)(1) = e^{n\varphi} \mathcal{S}_0(e^{2\varphi}h;n)(1),$$

i.e.,

$$c_{\frac{n}{2}} P_n(h)(\varphi) + \mathcal{S}(h;n)(1) = e^{n\varphi} \mathcal{S}(e^{2\varphi}h;n)(1). \tag{1.4.39}$$

This relation resembles the fundamental identity (1.3.7). In fact, the following result states that (1.4.39) and (1.3.7) are equivalent.

Theorem 1.4.3 ([GZ03]). *For even n,*

$$\mathcal{S}(h;n)(1) = (-1)^{\frac{n}{2}} c_{\frac{n}{2}} Q_n(h). \tag{1.4.40}$$

Thus, the Laurent series of the scattering operator $\mathcal{S}(h;\lambda)$ at $\lambda = n$ contains the critical GJMS-operator $P_n(h)$ *and* the critical Q-curvature $Q_n(h)$, and the fundamental identity for the pair (P_n, Q_n) is a consequence of the conformal transformation law of $\mathcal{S}$.

Next, we describe the consequences of these results for the families $\mathcal{T}_{2N}(h;\lambda)$ which define the sequence $a_0, a_2, a_4, \dots$ by

$$a_{2N}(h;\lambda) = \mathcal{T}_{2N}(h;\lambda)(a_0). \tag{1.4.41}$$

Proposition 1.4.3. *For $n \geq 3$, the family $\mathcal{T}_{2N}(h;\lambda)$ has the form*

$$\mathcal{T}_{2N}(h;\lambda) = \frac{1}{2^{2N}N!(\frac{n}{2}-\lambda-1)\cdots(\frac{n}{2}-\lambda-N)} P_{2N}(h;\lambda)$$

with a polynomial family $P_{2N}(h;\lambda)$ such that

$$P_{2N}(h;\lambda) = \Delta_h^N + LOT.$$

Moreover,

$$\operatorname{Res}_{\frac{n}{2}-N}(\mathcal{T}_{2N}(h;\lambda)) = -c_N P_{2N}(h) \tag{1.4.42}$$

and

$$P_{2N}\left(h;\frac{n}{2}-N\right) = P_{2N}(h). \tag{1.4.43}$$

Proof. We only describe the argument which yields (1.4.43) and (1.4.42). The cancellation of poles in both ladders yields

$$\operatorname{Res}_{\frac{n}{2}+N}(\mathcal{S}(h;\lambda)) + \operatorname{Res}_{\frac{n}{2}+N}(\mathcal{T}_{2N}(h;n-\lambda)) = 0$$

(see (1.4.19) for a special case). Hence

$$\operatorname{Res}_{\frac{n}{2}-N}(\mathcal{T}_{2N}(h;\lambda)) = \operatorname{Res}_{\frac{n}{2}+N}(\mathcal{S}(h;\lambda)).$$

Now Theorem 1.4.2 gives (1.4.42). On the other hand,

$$\operatorname{Res}_{\frac{n}{2}-N}(\mathcal{T}_{2N}(h;\lambda)) = -c_N P_{2N}\left(h;\frac{n}{2}-N\right).$$

This proves (1.4.43). □

In the following, it will be convenient to use the notation $\dot{f}(\lambda)$ for the derivative of a holomorphic function $f(\lambda)$. The above results imply a description of the critical Q-curvature $Q_n(h)$ in terms of the family $P_n(h;\lambda)$. This result will play a central role in Section 1.5.

Proposition 1.4.4. *For even $n \geq 2$,*

$$Q_n(h) = (-1)^{\frac{n}{2}} \dot{P}_n(h;0)(1).$$

Proof. The constant function $u = 1$ is harmonic ($\lambda = n$) and has a trivial asymptotic expansion. Hence

$$\lim_{\lambda \to 0} \mathcal{T}_n(h;\lambda)(1) = -\mathcal{S}(h;n)(1).$$

But we have

$$\mathcal{S}(h,n)(1) = -(-1)^{\frac{n}{2}} c_{\frac{n}{2}} Q_n(h)$$

by Theorem 1.4.3 and

$$\mathcal{T}_n(h;n)(1) = -c_{\frac{n}{2}} \dot{P}_n(h;0)(1)$$

by Proposition 1.4.3. This proves the assertion. □

We illustrate Proposition 1.4.4 by two examples.

Example 1.4.1.

$$\mathcal{T}_2(\lambda) = \frac{1}{2(n-2-2\lambda)}(\Delta - \lambda \mathsf{J}).$$

In particular,

$$P_2\left(\frac{n}{2}-1\right) = \Delta - \left(\frac{n}{2}-1\right)\mathsf{J} = P_2.$$

Example 1.4.2.

$$\begin{aligned}\mathcal{T}_4(\lambda) = \frac{1}{8(n-2-2\lambda)(n-4-2\lambda)}\big[&(\Delta-(\lambda+2)\mathsf{J})(\Delta-\lambda\mathsf{J}) \\ &+\lambda(2\lambda-n+2)|\mathsf{P}|^2 + 2(2\lambda-n+2)\delta(\mathsf{P}\#d) + (2\lambda-n+2)(d\mathsf{J},d)\big].\end{aligned}$$

In particular,

$$\begin{aligned}P_4\left(\frac{n}{2}-2\right)(u) = \left(\Delta-\frac{n}{2}\mathsf{J}\right)&\left(\Delta-\left(\frac{n}{2}-2\right)\mathsf{J}\right)(u) \\ &-(n-4)|\mathsf{P}|^2 u - 4\delta(\mathsf{P}\#du) - 2(d\mathsf{J},du).\end{aligned}$$

A calculation shows that this operator coincides with

$$\Delta^2 + \delta((n-2)\mathsf{J}g - 4\mathsf{P})\#d + \left(\frac{n}{2}-2\right)\left(\frac{n}{2}\mathsf{J}^2 - 2|\mathsf{P}|^2 - \Delta\mathsf{J}\right),$$

i.e., with the Paneitz operator P_4 (see (1.2.28)).

1.5 Residue families and the holographic formula for Q_n

In the present section, we introduce and discuss some one-parameter families of differential operators which will be called the *residue families*. The notation is motivated by the fact that they naturally arise from a certain residue construction.

Their relation to Q-curvatures and GJMS-operators resembles the properties of the scattering operator described in Section 1.4.

We emphasize that the discussion of the scattering operator in Section 1.4 ignored most of its deeper spectral theoretical properties. In fact, we were interested only in very rough information about its poles in $\frac{n}{2}+\mathbb{N}$ and certain related constructions. In particular, we completely ignored its global aspects, as the structure of resonances (poles in $\mathrm{Re}(s) < \frac{n}{2}$) and their relation to closed geodesics. With some exaggeration, one could say that we broke a fly on the wheel.

The residue families are local in nature and are designed to investigate the local structure of Q-curvature and GJMS-operators. They have additional properties which allow us to uncover the recursive structure of Q-curvatures and GJMS-operators. They naturally depend on a metric and are conformally covariant in a specific sense. Moreover, they are curved analogs of families of intertwining operators of principal series representations.

More precisely, for a given manifold (M, h) of dimension n, we define a sequence of families

$$D_{2N}^{res}(h;\lambda) : C^\infty(M \times [0,\varepsilon)) \to C^\infty(M), \; N \geq 1.$$

For odd n, the sequence is unbounded, but for even n, the families are defined only if $2N \leq n$. The main features of these families are the following. They

- are determined by and depend naturally on a metric h on M,
- satisfy a conformal transformation law,
- satisfy recursive relations.

Moreover, the properties of the critical family are closely related to the so-called *holographic formula* for the critical Q-curvature.

Let $g_+ = r^{-2}(dr^2 + h_r)$ be a Poincaré-Einstein metric with conformal infinity $[h]$, $h = h_0$. For odd n, we assume that h_r is even in r. For even n, the following constructions will only depend on the terms $h_0, h_{(2)}, \dots, h_{(n-2)}$ and $\mathrm{tr}\, h_{(n)}$. We recall that these are determined by h.

We start with the definition of the so-called holographic coefficients of h. Let $v(r) \in C^\infty(M)$ be defined as the quotient

$$v(r) = \frac{\mathrm{vol}(h_r)}{\mathrm{vol}(h)} \tag{1.5.1}$$

of volume forms. The Taylor series of $v(r)$ has the form

$$v(r) = 1 + v_2 r^2 + v_4 r^4 + \cdots + v_n r^n + \cdots \tag{1.5.2}$$

with coefficients $v_{2j} \in C^\infty(M)$, $j \geq 1$. For convenience, we set $v_0 = 1$. For odd n, h_r is uniquely determined (if even in r) by h. Thus, we have an *infinite* sequence v_{2j} of coefficients which are completely determined by h. More precisely, v_{2j} is given by a local formula which involves at most $2j$ derivatives of the metric. For even n, the situation is more subtle. In this case, only the *finite* sequence $v_2, \dots, v_n$ is uniquely determined by h. This is obvious for $v_2, \dots, v_{n-2}$. Although the coefficient v_n is influenced by $h_{(n)}$, which is *not* uniquely determined by h, v_n is a well-defined function of h since $h_{(n)}$ enters only through its uniquely determined trace. Moreover, v_{2j} is given by a local formula which involves at most $2j$ derivatives of the metric.

Definition 1.5.1 (Holographic coefficients). For even n, let $0 \le 2j \le n$, and for odd n, let $j \ge 0$. Then, the quantities $v_{2j}(h) \in C^\infty(M)$ are called the holographic coefficients of h.

Originally, the coefficients v_2, v_4 and v_6 were studied in connection with one of the early tests [HS98] of the AdS/CFT-duality [M98], [Wi98]. More precisely, the AdS/CFT-duality motivates the definition of a notion of volume of a space with an Einstein metric of negative curvature. Since such space have infinite volume, one is forced to apply an appropriate renormalization technique. This leads to the method of *holographic* renormalization [Sk02].

A general and rigorous mathematical discussion of these issues was given in [G00]. We briefly recall the main results. Let g_+ be a conformally compact Einstein metric on X with conformal infinity c on $M = \partial X$. The choice of $h \in c$ determines a boundary-defining function r so that $g_+ = r^{-2}(dr^2 + h_r)$ (near M) with $h_0 = h$. Hence

$$\operatorname{vol}(g_+) = r^{-(n+1)} \left(\operatorname{vol}(h_r)/\operatorname{vol}(h)\right) dr \operatorname{vol}(h).$$

We consider the asymptotics as $\varepsilon \to 0$ of the g_+-volume of the set $\{r > \varepsilon\}$. Using (1.5.2), we find

$$\int_{r>\varepsilon} \operatorname{vol}(g_+) = c_0 \varepsilon^{-n} + c_2 \varepsilon^{-(n-2)} + \cdots + c_{n-1} \varepsilon^{-1} + V + o(1)$$

for odd n, and

$$\int_{r>\varepsilon} \operatorname{vol}(g_+) = c_0 \varepsilon^{-n} + c_2 \varepsilon^{-(n-2)} + \cdots + c_{n-2} \varepsilon^{-2} - L \log \varepsilon + V + o(1)$$

for even n. The coefficients c_{2j} and L in these expansions are given by

$$c_{2j}(h) = \frac{1}{n-2j} \int_M v_{2j}(h) \operatorname{vol}(h) \tag{1.5.3}$$

and

$$L(h) = \int_M v_n(h) \operatorname{vol}(h). \tag{1.5.4}$$

Finally, the constant term $V(g_+, h)$ is called the *renormalized volume* of g_+ with respect to h.

Theorem 1.5.1 ([G00]). *For odd n, $V(g_+, h)$ does not depend on the choice of h. For even n, $L = L(g_+, h)$ does not depend on the choice of h.*

However, for even n, the renormalized volume $V(g_+, h)$ depends on the choice of h in the conformal infinity c of g_+. Its dependence on h can be described by the formula

$$(d/dt)|_0 \left(V(g_+, e^{2t\varphi}h)\right) = \int_M \varphi v_n(h)\,\mathrm{vol}(h) \tag{1.5.5}$$

Thus, v_n is the (infinitesimal) conformal anomaly of the renormalized volume of g_+. In the physical literature, v_n is often referred to as the *holographic anomaly*. (1.5.5) should be compared with Branson-Ørsted's *infinitesimal* Polyakov-type formula

$$(d/dt)|_0 \left(\log\det(D(e^{2t\varphi}g))\right) = \int_M \varphi a_n(D(g))\,\mathrm{vol}(g)$$

for the determinant of elliptic positive selfadjoint conformally covariant operators D. Here $a_n(D)$ is defined by the constant term in the expansion of the heat kernel of D. For details see [BO86], [BO88], [BO91a] and chapter 4 of [Br93].

For conformally compact Einstein metrics on four-manifolds, Anderson [A01] proved the relation

$$8\pi^2\chi(X) - 6V(g_+, h) = \int_{X^4} |\mathsf{W}(g_+)|^2\,\mathrm{vol}(g_+). \tag{1.5.6}$$

The integral on the right-hand side converges since on four-manifolds the integrand is invariant under conformal changes (see (1.2.18)). Anderson's formula (1.5.6) shows directly that $V(g_+, h)$ does not depend on h.

An extension of this formula to general odd dimensions n can be found in [CQY08]. Here the decomposition in Theorem 1.3.4 and the formula [FG02]

$$V(g_+, h) = -\int_M \dot{\mathcal{S}}(h; n)(1)\,\mathrm{vol}(h) \tag{1.5.7}$$

for the renormalized volume in terms of the scattering operator, play a key role. Using Proposition 1.4.2, (1.4.36) and $\mathcal{S}(h; n)(1) = 0$, the relation (1.5.7) re-proves that $V(g_+, h)$ does not depend on h.

We shall return to a discussion of the renormalized volume for *even* n at the end of Section 1.6.

The role of the coefficients v_{2j} in this context motivates us to refer to them as the holographic coefficients. The same convention was used in [J09b]. In the literature, one also finds the alternative notion of *renormalized volume coefficients* (see, for instance, [G09]).

Now we display explicit formulas for the first few holographic coefficients. Detailed proofs can be found in [J09b].

Example 1.5.1. In all dimensions $n \geq 2$,

$$v_2 = -\frac{1}{2}\mathsf{J} = -\frac{1}{2}\operatorname{tr}\mathsf{P}.$$

Example 1.5.2. In dimension $n \geq 3$,

$$v_4 = \frac{1}{8}(\mathsf{J}^2 - |\mathsf{P}|^2) = \frac{1}{4} \operatorname{tr} \wedge^2(\mathsf{P}).$$

Example 1.5.3. In dimension $4 \neq n \geq 3$,

$$v_6 = -\frac{1}{8} \operatorname{tr} \wedge^3(\mathsf{P}) - \frac{1}{24(n-4)}(\mathcal{B}, \mathsf{P})$$

with $\mathcal{B}$ as in (1.4.34).

Next, we recall that the families $\mathcal{T}_{2N}(h;\lambda)$ define the coefficients in the expansion

$$\sum_{N \geq 0} r^{\lambda+2N} \mathcal{T}_{2N}(h;\lambda)(f), \quad \mathcal{T}_0(h;\lambda)f = f \in C^\infty(M)$$

of a formal solution of the equation

$$-\Delta_{g_+} u = \lambda(n-\lambda)u$$

(see (1.4.41)). Now we have all ingredients to define residue families.

Definition 1.5.2 (Residue families). For even n and $2N \leq n$,

$$\begin{aligned} &D_{2N}^{res}(h;\lambda) \\ &= 2^{2N} N! \left[\left(-\frac{n}{2} - \lambda + 2N - 1\right) \cdots \left(-\frac{n}{2} - \lambda + N\right)\right] \delta_{2N}(h; \lambda + n - 2N) \end{aligned} \quad (1.5.8)$$

with

$$\begin{aligned} &\delta_{2N}(h;\lambda) \\ &= \sum_{j=0}^{N} \frac{1}{(2N-2j)!} \left[\mathcal{T}_{2j}^*(h;\lambda) \circ v_0 + \cdots + \mathcal{T}_0^*(h;\lambda) \circ v_{2j}\right] \circ i^*(\partial/\partial r)^{2N-2j}, \end{aligned} \quad (1.5.9)$$

where the embedding $i : M \hookrightarrow M \times [0,\varepsilon)$ is defined by $i(m) = (m,0)$, and the holographic coefficients are used as multiplication operators. $\mathcal{T}_{2j}^*(h;\lambda)$ denotes the formal adjoint of the operator $\mathcal{T}_{2j}(h;\lambda)$ on $C^\infty(M)$ with respect to the metric h. Note that $D_0^{res}(h;\lambda) = i^*$.

Later on, the critical family

$$\begin{aligned} D_n^{res}(h;\lambda) &= 2^n \left(\frac{n}{2}\right)! \left[\left(\frac{n}{2} - \lambda - 1\right) \cdots (-\lambda)\right] \\ &\times \sum_{j=0}^{\frac{n}{2}} \frac{1}{(n-2j)!} \left[\mathcal{T}_{2j}^*(h;\lambda) \circ v_0 + \cdots + \mathcal{T}_0^*(h;\lambda) \circ v_{2j}\right] \circ i^*(\partial/\partial r)^{n-2j} \end{aligned} \quad (1.5.10)$$

will be of special importance.

In connection with Definition 1.5.2, some comments are in order. First of all, the residue families $D_{2N}^{res}(h;\lambda)$ are completely determined by the given metric h. Although the residue families are linear combinations of compositions of differential operators on M and $M \times [0,\varepsilon)$ with the restriction i^*, we abuse language, and refer to them as differential operators. For generic λ, the definition of $D_{2N}^{res}(h;\lambda)$ involves $2N$ differentiations both along M and in the normal direction. In that sense, it is a family of differential operators of order $2N$. In particular, the critical residue family is of order n. Next, we recall that the families $\mathcal{T}_{2N}(h;\lambda)$ are rational in λ (see Proposition 1.4.3). This motivates the polynomial overall factor in (1.5.8). It has the effect that residue families are *polynomial* in λ. Although one can also define residue families for odd n, these will not be considered in what follows, and we omit their discussion.

Versions of residue families first appeared in the context of Selberg zeta functions (see the discussion in Chapter 1 of [J09b]). We expect that this relation serves as a source of similar constructions in other contexts.

For special values of the parameter λ, residue families degenerate to compositions of residue families of smaller order and GJMS-operators. The following observation is the simplest special cases of this effect.

Proposition 1.5.1. *For even n and $2N \le n$,*

$$D_{2N}^{res}\left(h;-\frac{n}{2}+N\right) = P_{2N}(h)i^*. \tag{1.5.11}$$

In particular,

$$D_n^{res}(h;0) = P_n(h)i^*, \tag{1.5.12}$$

i.e., the critical residue family at $\lambda = 0$ is given by the critical GJMS-operator.

Proof. The first statement in Proposition 1.4.3 implies that the only non-trivial contribution to $D_{2N}^{res}(h,-\frac{n}{2}+N)$ comes from

$$2^{2N}N!\left[\left(-\frac{n}{2}-\lambda+2N-1\right)\cdots\left(-\frac{n}{2}-\lambda+N\right)\right] \\ \times \mathcal{T}_{2N}^*(h;\lambda+n-2N) = P_{2N}^*(h;\lambda+n-2N)$$

at $\lambda = -\frac{n}{2}+N$. But (1.4.43) shows that

$$P_{2N}^*\left(h;\frac{n}{2}-N\right) = P_{2N}^*(h).$$

Now Corollary 1.4.1 completes the proof. □

The family $D_{2N}^{res}(h;\lambda)$ is the product of a polynomial of degree N in λ and the rational family $\delta_{2N}(h;\lambda)$. The polynomial factor has the effect of removing all poles. More precisely, we have

Proposition 1.5.2. *$D_{2N}^{res}(h;\lambda)$ is a polynomial of degree N in λ.*

Now we turn to a description of the conformal transformation law of residue families. We start with a discussion of the critical residue families. The formulation of the result requires one more ingredient. For conformally related metrics $\hat{h} = e^{2\varphi}h$, there exists a diffeomorphism κ which pulls back the corresponding Poincaré-Einstein metrics, i.e.,

$$\kappa^* \left(r^{-2}(dr^2 + h_r)\right) = r^{-2}(dr^2 + \hat{h}_r). \tag{1.5.13}$$

κ restricts to the identity map on the boundary. Thus, (1.5.13) implies

$$i^* \left(\frac{\kappa^*(r)}{r}\right)^2 \hat{h} = h,$$

i.e.,

$$i^* \left(\frac{\kappa^*(r)}{r}\right) = e^{-\varphi}. \tag{1.5.14}$$

Theorem 1.5.2. *For even n, the critical residue family satisfies the relations*

$$e^{-(\lambda-n)\varphi} \circ D_n^{res}(e^{2\varphi}h;\lambda) = D_n^{res}(h;\lambda) \circ \kappa_* \circ \left(\frac{\kappa^*(r)}{r}\right)^\lambda \tag{1.5.15}$$

for all $\varphi \in C^\infty(M)$.

The following result extends Theorem 1.5.2 to all residue families.

Theorem 1.5.3. *For even n and $2N \leq n$,*

$$e^{-(\lambda-2N)\varphi} \circ D_{2N}^{res}(e^{2\varphi}h;\lambda) = D_{2N}^{res}(h;\lambda) \circ \kappa_* \circ \left(\frac{\kappa^*(r)}{r}\right)^\lambda. \tag{1.5.16}$$

The following proof of Theorem 1.5.3 rests on an interpretation of $D_{2N}^{res}(h;\lambda)$ as a residue. In fact, it is this interpretation which motivates the terminology.

Proof. Let g be an asymptotically hyperbolic metric and ρ a boundary-defining function. Let

$$u \in \ker(\Delta_g + \mu(n-\mu))$$

with $\mathrm{Re}(\mu) = \frac{n}{2}$ and $\mu \neq \frac{n}{2}$ be an eigenfunction so that in

$$u = F\rho^\mu + G\rho^{n-\mu}$$

the functions F and G are smooth up to the boundary. Let $i^*(F) = f \in C^\infty(M)$. For sufficiently large $\mathrm{Re}(\lambda)$, we consider the function

$$\lambda \mapsto \int_X \rho^\lambda u \psi \,\mathrm{vol}(\rho^2 g) \tag{1.5.17}$$

for a test function $\psi \in C_0^\infty(\bar{X})$ with compact support up to the boundary. It admits a meromorphic continuation to $\mathbb{C}$ with *simple* poles in $\lambda = -\mu - 1 - N$, $N \in \mathbb{N}_0$ and respective residues

$$\int_M f\delta_N(\rho^2 g; \mu)(\psi) \operatorname{vol}(i^*(\rho^2 g)). \tag{1.5.18}$$

The differential operators $\delta_N(\cdot; \mu)$ depend on the metric $\rho^2 g$, i.e., on g and the defining function ρ. This follows by combining the asymptotic expansion of u with repeated partial integration.

In the setting of Poincaré-Einstein metrics, boundary-defining functions are determined by the choice of representing metrics in a conformal class. In this case, it is natural to regard δ_N as depending only on the metric on the boundary.

Let h and $\hat{h} = e^{2\varphi}h$, and let κ be the diffeomorphism which pulls back the corresponding Poincaré-Einstein metrics. Let u be an eigenfunction of the Laplacian of g_+ as above. In these terms, we consider the function

$$\lambda \mapsto \int_X r^\lambda \kappa^*(u)\psi \operatorname{vol}(r^2\kappa^*(g_+)). \tag{1.5.19}$$

$\kappa^*(u)$ is an eigenfunction of the Laplacian of $\kappa^*(g_+)$ with leading coefficient

$$i^*\left(\frac{\kappa^*(r)}{r}\right)^\mu f = e^{-\mu\varphi} f \in C^\infty(M)$$

(see (1.5.14)). Moreover, the residues of (1.5.19) are given by the formula

$$\int_M (e^{-\mu\varphi} f)\delta_N(\hat{h}; \mu)(\psi) \operatorname{vol}(\hat{h}) = \int_M e^{-(\mu-n)\varphi} f\delta_N(\hat{h}; \mu)(\psi) \operatorname{vol}(h). \tag{1.5.20}$$

On the other hand, for sufficiently large $\operatorname{Re}(\lambda)$, (1.5.19) equals

$$\begin{aligned}
&\int_X r^{\lambda+n+1}\kappa^*(u)\psi\kappa^*(r)^{-n-1}\kappa^*(\operatorname{vol}(dr^2 + h_r)) \\
&= \int_X \kappa_*(r)^{\lambda+n+1} u\kappa_*(\psi) r^{-n-1} \operatorname{vol}(dr^2 + h_r) \\
&= \int_X r^\lambda u \left(\frac{\kappa_*(r)}{r}\right)^{\lambda+n+1} \kappa_*(\psi) \operatorname{vol}(dr^2 + h_r).
\end{aligned}$$

Now regarding $(\kappa_*(r)/r)^{\lambda+n+1}\kappa_*(\psi)$ as a test function, we find that the residues are given by

$$\int_M f\delta_N(h; \mu)\left(\left(\frac{\kappa_*(r)}{r}\right)^{-\mu+n-N} \kappa_*(\psi)\right) \operatorname{vol}(h). \tag{1.5.21}$$

By uniqueness of analytic continuation, (1.5.20) and (1.5.21) coincide. Since $f \in C^\infty(M)$ is arbitrary, we find the relation

$$e^{-(n-\mu)\varphi} \circ \delta_N(\hat{h};\mu) = \delta_N(h;\mu) \circ \left(\frac{\kappa_*(r)}{r}\right)^{-\mu+n-N} \circ \kappa_*.$$

Now the substitution $\mu \mapsto \lambda+n-N$ yields the assertion. □

Theorem 1.5.2 has the following important consequence. We recall that $\dot{f}$ denotes the derivative of a holomorphic function f.

Corollary 1.5.1. *For even n and all $\varphi \in C^\infty(M)$,*

$$e^{n\varphi}\dot{D}_n^{res}(e^{2\varphi}h;0)(1) = \dot{D}_n^{res}(h;0)(1) - P_n(h)(\varphi). \tag{1.5.22}$$

Proof. We apply (1.5.15) to the constant function $u=1$, differentiate with respect to λ and set $\lambda=0$. Then we find

$$\begin{aligned} &-\varphi e^{n\varphi} D_n^{res}(\hat{h};0)(1) + e^{n\varphi}\dot{D}_n^{res}(\hat{h};0)(1) \\ &\qquad = \dot{D}_n^{res}(h;0)(\kappa_*(1)) + D_n^{res}(h;0)\left(\kappa_*\log\left(\frac{\kappa^*(r)}{r}\right)\right). \end{aligned}$$

Now using $\kappa_*(1)=1$ and $D_n^{res}(h;0) = P_n(h)i^*$ (see (1.5.12)), we obtain

$$\begin{aligned} e^{n\varphi}\dot{D}_n^{res}(\hat{h};0)(1) &= \dot{D}_n^{res}(h;0)(1) + P_n(h)\left(i^*\left(\kappa_*\log\left(\frac{\kappa^*(r)}{r}\right)\right)\right) \\ &= \dot{D}_n^{res}(h;0)(1) + P_n(h)\left(i^*\log\left(\frac{\kappa^*(r)}{r}\right)\right). \end{aligned}$$

Here we applied the fact that κ restricts to the identity on the boundary M. Now the relation (1.5.14) completes the proof. □

Now comparing (1.5.22) with the fundamental identity

$$e^{n\varphi}Q_n(e^{2\varphi}h) = Q_n(h) + (-1)^{\frac{n}{2}}P_n(h)(\varphi)$$

(see (1.3.7)) suggests that we ask whether it is true that $\dot{D}_n^{res}(h;0)(1) = -(-1)^{\frac{n}{2}} \cdot Q_n(h)$. The following result gives an affirmative answer. It will play an important role in Section 1.6.

Theorem 1.5.4. *On manifolds of even dimension n,*

$$\dot{D}_n^{res}(h;0)(1) = -(-1)^{\frac{n}{2}}Q_n(h). \tag{1.5.23}$$

(1.5.23) together with $D_n^{res}(h;0) = P_n(h)i^*$ (see (1.5.12)) should be compared with

$$\mathcal{S}(h;n)(1) = (-1)^{\frac{n}{2}}c_{\frac{n}{2}}Q_n(h) \quad \text{and} \quad \operatorname{Res}_n(\mathcal{S}(h;\lambda)) = -c_{\frac{n}{2}}P_n(h).$$

The proof of Theorem 1.5.4 will be given below. It is closely connected with the proof of another formula for the critical Q-curvature which sometimes is called the *holographic formula.*

Theorem 1.5.5 (The holographic formula). *On manifolds of even dimension n,*

$$(-1)^{\frac{n}{2}} n Q_n(h) = 2^{n-1}\left(\frac{n}{2}\right)!\left(\frac{n}{2}-1\right)!\sum_{j=0}^{\frac{n}{2}-1}(n-2j)\mathcal{T}_{2j}^*(h;0)(v_{n-2j}(h)). \quad (1.5.24)$$

The detailed proof of the holographic formula for the critical Q-curvature can be found in [GJ07]. The arguments rest on results in [GZ03] and their subsequent refinements in [FG02]. See also the description in [J09b].

In connection with (1.5.24), the adjective "holographic" is motivated by the recent role of holography in physics. It is well known that holography is a technical device to encode a three-dimensional object on a two-dimensional medium. In theoretical physics, the notion of holography has been used during the last decade in connection with the proposed AdS/CFT-duality between gauge field theory in dimension 4 and gravity in dimension (at least) 5. For more details we refer to the foundational works [M98], [Wi98] and the reviews [AGMOO00], [HF04].

(1.5.24) relates the Riemannian curvature quantity $Q_n(h)$ of the manifold (M, h) to data which are constructed in terms of an Einstein metric on a space of *one higher* dimension. It is this aspect which is stressed by referring to it as the holographic formula.

In small dimensions, the holographic formula can be made more explicit. We display the first three cases.

Example 1.5.4. In dimension $n = 2$, $Q_2 = -2v_2$. This follows from $Q_2 = \mathrm{scal}/2$ and $v_2 = -\mathsf{J}/2$ (see Example 1.5.1).

Note that the formula $Q_2 = -v_2$ is valid in all dimensions.

Example 1.5.5. In dimension $n = 4$,

$$Q_4 = 16v_4 + 8\mathcal{T}_2^*(0)(v_2)$$

with

$$v_2 = -\frac{1}{2}\mathsf{J} \quad \text{and} \quad v_4 = \frac{1}{4}\operatorname{tr}\wedge^2(\mathsf{P}) = \frac{1}{8}(\mathsf{J}^2 - |\mathsf{P}|^2)$$

(see Example 1.5.2). By Example 1.4.1, this formula is equivalent to

$$Q_4 = 16v_4 + 2P_2^*(0)(v_2) = 16v_4 + 2\Delta(v_2).$$

The formula in Example 1.5.5 does *not* hold in dimension $n \neq 4$.

Example 1.5.6. In dimension $n = 6$,

$$Q_6 = -384v_6 - 32P_2^*(0)(v_4) - 2P_4^*(0)(v_2).$$

This formula is equivalent to

$$Q_6 = -384v_6 + \Delta^2\mathsf{J} - 8\delta(\mathsf{P}\#d\mathsf{J}) + 4\Delta(|\mathsf{P}|^2 - \mathsf{J}^2),$$

where

$$v_6 = -\frac{1}{8}\operatorname{tr}\wedge^3(\mathsf{P}) - \frac{1}{48}(\mathcal{B},\mathsf{P}) \tag{1.5.25}$$

(see Example 1.5.3).

The formula for Q_6 in Example 1.5.6 does *not* hold in dimension $n \neq 6$.

The following conjecture extends the holographic formula to subcritical Q-curvatures in even dimension.

Conjecture 1.5.1. *For even n and $2N \leq n$,*

$$4Nc_N Q_{2N}(h) = \sum_{j=0}^{N-1}(2N-2j)\mathcal{T}_{2j}^*\left(h;\frac{n}{2}-N\right)(v_{2N-2j}(h)), \tag{1.5.26}$$

where $c_N = (-1)^N(2^{2N}N!(N-1)!)^{-1}$.

For more details around this conjecture we refer to [J09b].

Now we discuss two general consequences of the holographic formula. The following result is due to Graham and Zworski [GZ03].

Corollary 1.5.2. *For closed manifolds M of even dimension n, we have the relation*

$$\int_M Q_n(h)\operatorname{vol}(h) = 2^{n-1}(-1)^{\frac{n}{2}}\left(\frac{n}{2}\right)!\left(\frac{n}{2}-1\right)!\int_M v_n(h)\operatorname{vol}(h) \tag{1.5.27}$$

of conformal invariants.

Proof. Splitting off the term with v_n in (1.5.24) yields

$$(-1)^{\frac{n}{2}} nQ_n = 2^{n-1}\left(\frac{n}{2}\right)!\left(\frac{n}{2}-1\right)!\left(nv_n + \sum_{j=1}^{\frac{n}{2}-1}(n-2j)\mathcal{T}_{2j}^*(0)(v_{n-2j})\right).$$

But the terms in the sum integrate to 0. In fact, since the harmonic function $u = 1$ has a trivial asymptotic expansion, we have $\mathcal{T}_{2j}(0)(1) = 0$ for $j \geq 1$. It follows that

$$\int_M \mathcal{T}_{2j}^*(h;0)(v_{n-2j}(h))\operatorname{vol}(h) = \int_M v_{n-2j}(h)\mathcal{T}_{2j}(h;0)(1)\operatorname{vol}(h) = 0.$$

This proves the assertion. □

By Theorem 1.3.4, the conformal invariance of both sides of (1.5.27) implies that Q_n and v_n both can be written as linear combinations of the Chern-Gauß-Bonnet integrand, i.e., a multiple of the Pfaffian form, a local conformal invariant, and a divergence. In the locally conformally flat case, it already follows from [BGP95] that Q_n and v_n are linear combinations of the Pfaffian and a divergence. But for conformally flat metrics, the holographic anomaly v_n actually is just a multiple of the Pfaffian. Therefore, the holographic formula for Q_n implies an explicit decomposition of Q_n as a sum of a multiple of the Pfaffian and a certain divergence. The problem of finding a direct link between Q_n and the Pfaffian has been formulated at various places in the literature [ES03].

In order to precisely formulate the result, we first recall the definition of the Pfaffian form and its relation to the Chern-Gauß-Bonnet formula. We define the Pfaffian form by

$$\mathrm{Pf}_n = \frac{1}{2^{\frac{n}{2}}(\frac{n}{2})!}\sum_{\sigma\in S_n}\epsilon(\sigma)\Omega_{\sigma_1\sigma_2}\wedge\cdots\wedge\Omega_{\sigma_{n-1}\sigma_n}\in\Omega^n(M),$$

where Ω_{ij} are the curvature forms $\Omega_{ij}(\cdot,\cdot) = g(R(\cdot,\cdot)e_i,e_j)$ with respect to a local orthonormal basis $\{e_i\}$. Moreover, let

$$E_n = (-2\pi)^{-\frac{n}{2}}\,\mathrm{Pf}_n$$

be the Euler form. In these terms, the Chern-Gauß-Bonnet theorem [G04] states that

$$\int_{M^n} E_n = \chi(M^n). \tag{1.5.28}$$

In the following, we identify E_n and Pf_n with its respective density with respect to the Riemannian volume form. In particular, we have

$$\mathrm{Pf}_2 = R_{1212} = -K = -\mathsf{J}$$

and

$$\begin{aligned}8\,\mathrm{Pf}_4 &= \sum_{i,j,k,l}(R_{ijkl}R_{ijkl} - 4R_{ijik}R_{ljlk} + R_{ijij}R_{klkl})\\ &= |R|^2 - 4|\,\mathrm{Ric}\,|^2 + \mathrm{scal}^2\\ &= |\mathsf{W}|^2 + 8(\mathsf{J}^2 - |\mathsf{P}|^2).\end{aligned}$$

Corollary 1.5.3. *On manifolds of even dimension n with a locally conformally flat metric h, the critical Q-curvature naturally decomposes in the form*

$$\begin{aligned}(-1)^{\frac{n}{2}}Q_n(h) &= [(n-2)(n-4)\cdots 2]\,\mathrm{Pf}_n\\ &\quad + 2^{n-2}\Big(\frac{n}{2}-1\Big)!\Big(\frac{n}{2}-1\Big)!\sum_{j=1}^{\frac{n}{2}-1}(n-2j)\mathcal{T}_{2j}^*(h;0)(v_{n-2j}(h)).\end{aligned} \tag{1.5.29}$$

The sum on the right-hand side is a divergence. [2]

Note that (1.5.29) is a concrete version of the decomposition given by Theorem 1.3.4.

[2] In [GJ07], the formula corresponding to (1.5.29) does not contain the coefficient $(-1)^{\frac{n}{2}}$. The difference is due to the different sign convention for the curvature tensor.

Proof. The proof of Corollary 1.5.2 yields the decomposition (1.5.29), up to the identification of

$$2^{n-1}\left(\frac{n}{2}\right)!\left(\frac{n}{2}-1\right)!v_n$$

with

$$[(n-2)(n-4)\cdots 2]\,\mathrm{Pf}_n\,.$$

In order to prove that relation, we apply the result that, under the assumption $\mathsf{W}=0$, the holographic coefficients are given by

$$v_{2j}=(-2)^{-j}\operatorname{tr}\wedge^j(\mathsf{P}) \tag{1.5.30}$$

(see [GJ07], [FG07]). In particular,

$$v_n=(-2)^{-\frac{n}{2}}\operatorname{tr}\wedge^{\frac{n}{2}}(\mathsf{P}).$$

On the other hand, $\mathsf{W}=0$ yields $R=-\mathsf{P}\owedge h$, which in turn gives

$$\mathrm{Pf}_n=(-1)^{\frac{n}{2}}\left(\frac{n}{2}\right)!\operatorname{tr}\wedge^{\frac{n}{2}}(\mathsf{P}).$$

These results yield

$$2^{\frac{n}{2}}\left(\frac{n}{2}\right)!v_n=\mathrm{Pf}_n\,.$$

The proof is complete. □

Example 1.5.7. Let $n=2$. Then $Q_2=-\mathrm{Pf}_2=K$.

The following example was already mentioned at the end of Section 1.2.

Example 1.5.8. Let $n=4$ and $\mathsf{W}=0$. Then

$$Q_4=2\,\mathrm{Pf}_4+8\mathcal{T}_2^*(0)(v_2)=2\,\mathrm{Pf}_4-\Delta\mathsf{J}$$

by using $v_2=-\mathsf{J}/2$ and Example 1.4.1. This formula is a special case of (1.2.44).

Further consequences of the holographic formula (or rather Theorem 1.5.4) will be described in Section 1.6.

We continue with a *proof of Theorem* 1.5.4.

First of all, a direct evaluation of $D_n^{res}(h;0)(1)$ yields the following intermediate result.

Proposition 1.5.3. *For even* n,

$$\dot{D}_n^{res}(h;0)(1)=-2^n\left(\frac{n}{2}\right)!\left(\frac{n}{2}-1\right)!\sum_{j=0}^{\frac{n}{2}-1}\mathcal{T}_{2j}^*(h;0)(v_{n-2j}(h))+\dot{P}_n^*(h;0)(1).$$

Next, we recall that

$$\dot{P}_n(h;0)(1) = (-1)^{\frac{n}{2}} Q_n(h) \tag{1.5.31}$$

by Proposition 1.4.4. In contrast to $P_n(h;0)$, the derivative $\dot{P}_n(h;0)$ is not self-adjoint. However, we have the following identity.

Theorem 1.5.6 ([GJ07]). *For even n,*

$$n\left(\dot{P}_n(h;0) - \dot{P}_n^*(h;0)\right)(1) = -2^n \left(\frac{n}{2}\right)! \left(\frac{n}{2}-1\right)! \sum_{j=1}^{\frac{n}{2}-1} 2j\mathcal{T}_{2j}^*(h;0)(v_{n-2j}(h)).$$

Proposition 1.5.3 and (1.5.31) imply

$$\begin{aligned} n\dot{D}_n^{res}(h;0)(1) = -2^n \left(\frac{n}{2}\right)! \left(\frac{n}{2}-1\right)! \sum_{j=0}^{\frac{n}{2}-1} n\mathcal{T}_{2j}^*(h;0)(v_{n-2j}(h)) \\ + n\left(\dot{P}_n^*(h;0) - \dot{P}_n(h;0)\right)(1) + n(-1)^{\frac{n}{2}} Q_n(h). \end{aligned}$$

Thus, by Theorem 1.5.6, the quantity $n\dot{D}_n^{res}(h;0)(1)$ equals

$$-2^n \left(\frac{n}{2}\right)! \left(\frac{n}{2}-1\right)! \sum_{j=0}^{\frac{n}{2}-1} (n-2j)\mathcal{T}_{2j}^*(h;0)(v_{n-2j}(h)) + n(-1)^{\frac{n}{2}} Q_n(h).$$

Now the holographic formula Theorem 1.5.5 yields

$$n\dot{D}_n^{res}(h;0)(1) = -(-1)^{\frac{n}{2}} 2nQ_n(h) + n(-1)^{\frac{n}{2}} Q_n(h) = -(-1)^{\frac{n}{2}} nQ_n(h).$$

This proves Theorem 1.5.4.

1.6 Recursive structures

In the present section, we discuss the recursive structure of residue families, and demonstrate how it can be used to derive recursive formulas for both Q-curvatures and GJMS-operators.

The residue families are recursive in the sense that any given family can be expressed (even in various ways) in terms of lower-order residue families *and* GJMS-operators.

For the flat metric on $\mathbb{R}^n$, residue families can be interpreted as intertwining families for principal series representations. From that point of view, their factorization identities are consequences of multiplicity-free branching laws for Verma modules. For full details concerning this aspect we refer to [J09b].

In view of Theorem 1.5.3, factorization identities for residue families of the flat metric imply factorization identities for residue families for (locally) conformally flat metrics. More generally, we *conjecture* that the validity of factorization identities extends to general metrics in full generality. We discuss some results which support this conjecture.

The residue families give rise to the notion of Q-curvature polynomials (or Q-polynomials for short). These are polynomials of one variable which generalize Q-curvature. The fact that the Q-polynomials are recursively determined is the key to uncover recursive formulas for Q-curvatures in terms of lower-order Q-curvatures and lower-order GJMS-operators. In this context, we shall often restrict to locally conformally flat metrics. More general results follow as soon as the appropriate factorization identities are available.

We continue with a description of the details. Full proofs can be found in [J09b], [FJ] and [J09a].

We start with formulation of the factorization identities for the residue families of the flat metric h_c on $\mathbb{R}^n$.

Theorem 1.6.1. *The family*

$$D_{2N}^{res}(h_c;\lambda): C^\infty(\mathbb{R}^{n+1}) \to C^\infty(\mathbb{R}^n)$$

factorizes for the $2N+2$ *parameters*

$$\lambda \in \left\{-\frac{n}{2}+N,\dots,-\frac{n}{2}+2N\right\} \cup \left\{-\frac{n+1}{2},\dots,-\frac{n+1}{2}+N\right\}$$

as

$$D_{2N}^{res}\left(h_c;-\frac{n}{2}+2N-j\right) = P_{2j}(h_c)\circ D_{2N-2j}^{res}\left(h_c;-\frac{n}{2}+2N-j\right) \tag{1.6.1}$$

for $j=0,\dots,N$ *and*

$$D_{2N}^{res}\left(h_c;-\frac{n+1}{2}+j\right) = D_{2N-2j}^{res}\left(h_c;-\frac{n+1}{2}-j\right)\circ P_{2j}(dr^2+h_c) \tag{1.6.2}$$

for $j=0,\dots,N$. *Here*

$$P_{2j}(h_c) = \Delta_{\mathbb{R}^n}^j \quad \textit{and} \quad P_{2j}(dr^2+h_c) = \Delta_{\mathbb{R}^{n+1}}^j$$

are the GJMS-operators on $\mathbb{R}^n$ *and* $\mathbb{R}^{n+1}$.

Besides the two trivial relations for $j=0$, the relations in (1.6.1) and (1.6.2) contain $2N$ non-trivial identities. We recall from Section 1.5 that $D_{2N}^{res}(h;\lambda)$ is a polynomial of degree N in λ. Thus, Theorem 1.6.1 implies that the $N+1$ coefficients of $D_{2N}^{res}(h_c;\lambda)$ can be written in terms of lower-order residue families for h_c and powers of the Laplacian on $\mathbb{R}^n$ and $\mathbb{R}^{n+1}$. Applying this argument to the lower-order residue families, finally yields a formula for the family $D_{2N}^{res}(h_c;\lambda)$ in terms of powers of Laplacians on $\mathbb{R}^n$ and $\mathbb{R}^{n+1}$. Since for $N \geq 2$, the number $2N$ of non-trivial relations for $D_{2N}^{res}(h_c;\lambda)$ exceeds the number $N+1$ of operator coefficients, the form of the resulting formulas depends on the choices of factorization identities used in this process. One of the natural choices consists in using the system (1.6.1) together with the identity for $j=1$ of (1.6.2). This choice will play a special role in what follows.

In order to illustrate the situation, we display two examples of low orders.

Example 1.6.1. The definition of $D_2^{res}(h_c;\lambda)$ yields

$$-(2\lambda+n-2)i^*(\partial/\partial r)^2 + \Delta_{\mathbb{R}^n} i^*.$$

This formula can be rewritten as

$$D_2^{res}(h_c;\lambda) = -(2\lambda+n-2)i^*\Delta_{\mathbb{R}^{n+1}} + (2\lambda+n-1)\Delta_{\mathbb{R}^n} i^*.$$

For the parameters $\lambda \in \left\{-\frac{n}{2}+1, -\frac{n+1}{2}+1\right\}$, the family specializes to

$$\Delta_{\mathbb{R}^n} i^* \quad \text{and} \quad i^*\Delta_{\mathbb{R}^{n+1}},$$

respectively. These are the two non-trivial factorization identities. These identities characterize the family which is linear in λ.

Example 1.6.2. We consider the critical residue family of order 4 for the flat metric h_c on $\mathbb{R}^4$. The definition of $D_4^{res}(h_c;\lambda)$ yields

$$\frac{4}{3}\lambda(\lambda-1)i^*(\partial/\partial r)^4 - 4\lambda\Delta_{\mathbb{R}^4} i^*(\partial/\partial r)^2 + \Delta_{\mathbb{R}^4}^2 i^*. \tag{1.6.3}$$

This family is quadratic in λ and satisfies four non-trivial factorization identities:

$$\begin{aligned} D_4^{res}(h_c;0) &= \Delta_{\mathbb{R}^4}^2 i^*, \\ D_4^{res}(h_c;1) &= \Delta_{\mathbb{R}^4} D_2^{res}(h_c;1), \\ D_4^{res}(h_c;-3/2) &= D_2^{res}(h_c;-7/2)\Delta_{\mathbb{R}^5}, \\ D_4^{res}(h_c;-1/2) &= i^*\Delta_{\mathbb{R}^5}^2. \end{aligned}$$

The following results are curved analogs of Example 1.6.1 and Example 1.6.2.

Theorem 1.6.2. *For any manifold (M,h) of dimension $n \geq 2$, the residue family of order 2 is given by*

$$D_2^{res}(h;\lambda) = -(2\lambda+n-2)i^*(\partial/\partial r)^2 + (\Delta + \lambda \mathsf{J})i^*,$$

where Δ and J are to be understood with respect to h. It satisfies the identities

$$D_2^{res}\left(h;-\frac{n}{2}+1\right) = P_2(h)i^* \quad \textit{and} \quad D_2^{res}\left(h;-\frac{n+1}{2}+1\right) = i^*P_2(dr^2+h_r).$$

Theorem 1.6.3. *For any four-manifold (M, h), the critical residue family of order 4 is given by*

$$D_4^{res}(h;\lambda) = \frac{4}{3}\lambda(\lambda-1)i^*(\partial/\partial r)^4 - 4\lambda(\Delta + (\lambda-2)\mathsf{J})i^*(\partial/\partial r)^2 \\ + \left[P_4^*(\lambda) + 4\lambda(\Delta - \lambda\mathsf{J})\mathsf{J} + 4\lambda(\lambda-1)(\mathsf{J}^2 - |\mathsf{P}|^2)\right] i^*$$

with

$$P_4(\lambda) = (\Delta - (\lambda+2)\mathsf{J})(\Delta - \lambda\mathsf{J}) \\ + 2\lambda(\lambda-1)|\mathsf{P}|^2 + 4(\lambda-1)\delta(\mathsf{P}\#d) + 2(\lambda-1)(d\mathsf{J}, d). \quad (1.6.4)$$

Here Δ, P, J are to be understood with respect to h. It satisfies the four factorization identities

$$D_4^{res}(h;0) = P_4(h)i^*,$$
$$D_4^{res}(h;1) = P_2(h)D_2^{res}(h;1),$$
$$D_4^{res}(h;-3/2) = D_2^{res}(h;-7/2)P_2(dr^2 + h_r),$$
$$D_4^{res}(h;-1/2) = i^*P_4(dr^2 + h_r).$$

These results for general metrics can be proved by direct calculations, i.e., by using the definitions of residue families in terms of the operators $\mathcal{T}_{2j}$ and the holographic coefficients v_{2j}. However, for locally conformally flat metrics h, the factorization identities in Theorem 1.6.2 and Theorem 1.6.3 also follow from those in the flat case by appealing to Theorem 1.5.3. More generally, for such metrics, these results are special cases of the following curved version of Theorem 1.6.1. We note that for locally conformally flat h, the Taylor series of h_r breaks off and

$$h_r = h - r^2\mathsf{P} + \frac{r^4}{4}\mathsf{P}^2. \quad (1.6.5)$$

Moreover, for such h, there are GJMS-operators to all orders [FG07].

Theorem 1.6.4. *Assume that (M, h) is locally conformally flat. Then, the family*

$$D_{2N}^{res}(h;\lambda) : C^\infty(M \times [0,\varepsilon)) \to C^\infty(M)$$

factorizes for the $2N+2$ parameters

$$\lambda \in \left\{-\frac{n}{2}+N, \dots, -\frac{n}{2}+2N\right\} \cup \left\{-\frac{n+1}{2}, \dots, -\frac{n+1}{2}+N\right\}$$

as

$$D_{2N}^{res}\left(h; -\frac{n}{2}+2N-j\right) = P_{2j}(h) \circ D_{2N-2j}^{res}\left(h; -\frac{n}{2}+2N-j\right) \quad (1.6.6)$$

for $j = 0, \dots, N$ *and*

$$D_{2N}^{res}\left(h; -\frac{n+1}{2}+j\right) = D_{2N-2j}^{res}\left(h; -\frac{n+1}{2}-j\right) \circ P_{2j}(dr^2+h_r) \qquad (1.6.7)$$

for $j = 0, \dots, N$. *Here* $P_{2j}(h)$ *and* $P_{2j}(dr^2+h_r)$ *are the GJMS-operators of* h *and the conformal compactification* $dr^2 + h_r$ *of the corresponding Poincaré-Einstein metric.*

In the following, it will be convenient to use the notation

$$\bar{P}_2(h) = P_2(dr^2 + h_r) \quad \text{and} \quad \bar{Q}_2(h) = Q_2(dr^2 + h_r). \qquad (1.6.8)$$

Theorem 1.6.4 was formulated under the assumption that h is locally conformally flat. The following conjecture states that this assumption can be removed.

Conjecture 1.6.1. *For even* n *and* $2N \leq n$, *all* $2N$ *non-trivial factorization identities in* (1.6.6) *and* (1.6.7) *hold true for any metric* h.

Theorem 1.6.2 and Theorem 1.6.3 provide support for Conjecture 1.6.1. Next, we stress that the nature of the factorization identities in the system (1.6.6) differs from the nature of those in the system (1.6.7). In fact, while the identities in (1.6.6) contain a GJMS-operator of the metric h as a factor, the identities (1.6.7) contain a factor which is a GJMS-operator of the metric $dr^2 + h_r$. This difference is also reflected by the fact that the validity of the system (1.6.6) can be proved for all h (see Section 6.12 in [J09b]), whereas no such general result is known for the second system. The following discussion will show that already the validity of one of the factorization identities in the system (1.6.7), namely that which contains the Yamabe operator $\bar{P}_2$, would have interesting consequences for Q-curvature. For more details see also [FJ].

Now we turn to the discussion of consequences of factorization identities for Q-curvatures. First, we show that Theorem 1.6.3 implies the following recursive formula for the critical Q-curvature Q_4.

Proposition 1.6.1. *For any four-manifold* (M, h),

$$Q_4 = P_2(Q_2) - 2i^* \bar{P}_2(\bar{Q}_2). \qquad (1.6.9)$$

Proof. We combine the formula

$$\dot{D}_4^{res}(h; 0)(1) = -Q_4(h) \qquad (1.6.10)$$

(see Theorem 1.5.4) with the factorization identities

$$D_4^{res}(h; 1)(1) = P_2(h) D_2^{res}(h; 1)(1),$$
$$D_4^{res}(h; -3/2)(1) = D_2^{res}(h; -7/2) \bar{P}_2(h)(1)$$

following from Theorem 1.6.3. By Theorem 1.6.2, the latter two relations are equivalent to

$$D_4^{res}(h;1)(1) = P_2(h)(-4i^*\bar{P}_2(h) + 5P_2(h)i^*)(1),$$
$$D_4^{res}(h;-3/2)(1) = (5i^*\bar{P}_2(h) - 4P_2(h)i^*)\bar{P}_2(h)(1).$$

Now $D_4^{res}(h;\lambda)(1)$ is a quadratic polynomial in λ with vanishing constant term, i.e.,

$$D_4^{res}(h;\lambda)(1) = \alpha\lambda^2 + \beta\lambda.$$

In these terms, the above relations read

$$\begin{pmatrix} \frac{9}{4} & -\frac{3}{2} \\ 1 & 1 \end{pmatrix} \begin{pmatrix} \alpha \\ \beta \end{pmatrix} = \begin{pmatrix} (5i^*\bar{P}_2(h) - 4P_2(h)i^*)\bar{P}_2(h)(1) \\ P_2(h)(-4i^*\bar{P}_2(h) + 5P_2(h)i^*)(1) \end{pmatrix}. \tag{1.6.11}$$

Solving the matrix equation yields

$$\beta = -P_2(Q_2) + 2i^*\bar{P}_2(\bar{Q}_2).$$

Here we have used $P_2(1) = -Q_2$, $\bar{P}_2(1) = -\frac{3}{2}\bar{Q}_2$ and the restriction property

$$i^*\bar{Q}_2 = Q_2. \tag{1.6.12}$$

But (1.6.10) gives $\beta = -Q_4(h)$. The proof is complete. □

Proposition 1.6.1 acts as a recursive formula for Q_4. In fact, it expresses the fourth-order curvature quantity Q_4 in terms of second-order constructions: the Yamabe operator and Q_2 for the given metric h and $dr^2 + h_r$.

It is remarkable that the formula for Q_4 in Proposition 1.6.1 literally holds true in all dimensions $n \geq 3$. We refer to this result as *universality*. Indeed, in dimension $n \geq 3$, the assertion reads

$$Q_4 = \left(\Delta - \frac{n-2}{2}\mathsf{J}\right)(\mathsf{J}) - 2i^*\left((\partial/\partial r)^2 + \Delta_{h_r} - \frac{n-1}{2}\bar{Q}_2\right)(\bar{Q}_2).$$

Now using $i^*\bar{Q}_2 = Q_2 = \mathsf{J}$ (see (1.6.12)) and $i^*(\partial/\partial r)^2(\bar{Q}_2) = |\mathsf{P}|^2$, the sum simplifies to

$$\frac{n}{2}\mathsf{J}^2 - 2|\mathsf{P}|^2 - \Delta\mathsf{J}.$$

This proves the universality of (1.6.1).

The following result is an analog for Q_6 of Proposition 1.6.1.

Proposition 1.6.2. *The critical Q-curvature Q_6 satisfies the recursive formula*

$$Q_6 = \frac{2}{3}(P_2(Q_4) + P_4(Q_2)) - \frac{5}{3}P_2^2(Q_2) + \frac{8}{3}i^*\bar{P}_2^2(\bar{Q}_2). \tag{1.6.13}$$

Moreover, (1.6.13) *is universal, i.e., holds true in all dimensions $n \geq 6$.*

We outline the method of the proof of Proposition 1.6.2. We first restrict to a locally conformally flat metric h. The residue family $D_6^{res}(h;\lambda)$ is a cubic polynomial in λ which satisfies six factorization identities. Three of these identities involve the GJMS-operators P_2, P_4 and P_6 for h. Hence the family $D_6^{res}(h;\lambda)$ can be written as a linear combination of the right-hand sides of these identities and the right-hand side of the identity which involves the Yamabe operator $\bar{P}_2(h)$. In turn, the lower-order residue families $D_2^{res}(h;\lambda)$ and $D_4^{res}(h;\lambda)$ which appear in the resulting formula can be written as linear combinations of the corresponding right-hand sides of their factorization relations. This leads to a formula for $D_6^{res}(h;\lambda)$ as a linear combination of compositions of the GJMS-operators $P_6(h)$, $P_4(h)$, $P_2(h)$ and the Yamabe operator $\bar{P}_2(h) = P_2(dr^2+h_r)$. Now combining the resulting formula with

$$\dot{D}_6^{res}(h;0)(1) = Q_6(h)$$

(see Theorem 1.5.4) yields a formula for Q_6 as a linear combination of compositions of the GJMS-operators $P_2(h)$, $P_4(h)$ and the Yamabe operator $\bar{P}_2(h) = P_2(dr^2 + h_r)$ (acting on $u = 1$). That formula contains compositions of GJMS-operators for h with powers of $\bar{P}_2(h)$ up to 3. Next, we express the quantities

$$i^*\bar{P}_2(h)(1) \quad \text{and} \quad i^*\bar{P}_2^2(h)(1)$$

in terms of subcritical GJMS-operators and subcritical Q-curvatures. For that purpose, we apply the formulas

$$i^*\bar{Q}_2 = Q_2 \quad \text{and} \quad -2i^*\bar{P}_2(\bar{Q}_2) = Q_4 - P_2(Q_2) \tag{1.6.14}$$

(see (1.6.12) and (1.6.9)) in dimension $n = 6$. Here it is crucial that the relations in (1.6.14) hold true in dimension $n = 6$. This is a special case of their universality. The final result is (1.6.13).

A direct proof of (1.6.13) for general metrics in dimension $n = 6$ as well as a proof of the universality of (1.6.13) can be given by making both sides explicit. An alternative proof for general metrics in the critical dimension follows from an extension of the appropriate factorization identities. For the full details we refer to [J09b].

An extension of these methods yields the following recursive formula for the critical Q-curvature Q_8.

Proposition 1.6.3. *For locally conformally flat metrics, the critical Q-curvature Q_8 satisfies the recursive formula*

$$\begin{aligned} Q_8 = \frac{3}{5}P_2(Q_6) &+ \left[-4P_2^2 + \frac{17}{5}P_4\right](Q_4) \\ &+ \left[-\frac{22}{5}P_2^3 + \frac{8}{5}P_2P_4 + \frac{28}{5}P_4P_2 - \frac{9}{5}P_6\right](Q_2) - \frac{16}{5}i^*\bar{P}_2^3(\bar{Q}_2). \end{aligned} \tag{1.6.15}$$

The recursive formulas in Propositions 1.6.1–1.6.3 express Q_4, Q_6 and Q_8 in terms of respective lower-order GJMS-operators, lower-order Q-curvatures, and a respective term of the form

$$i^* \bar{P}_2^N(\bar{Q}_2), \quad N = 1, 2, 3. \tag{1.6.16}$$

In contrast to the holographic formula in Theorem 1.5.5 and its extension in Conjecture 1.5.1, the terms (1.6.16) are the only ones which are defined through the associated Poincaré-Einstein metric. In the following, we shall discuss another type of recursive representations for Q-curvature which mixes features of the holographic formula and the above recursive formulas: they relate Q_{2N} to v_{2N} (as in (1.5.26)), up to lower-order GJMS-operators and Q-curvatures. These formulas arise through the theory of Q-curvature polynomials (or Q-polynomials, for short).

Definition 1.6.1 (Q-curvature polynomials). For even $n \geq 2$ and $1 \leq N \leq \frac{n}{2}$, the degree N polynomial

$$Q_{2N}^{res}(h; \lambda) = -(-1)^N D_{2N}^{res}(h; \lambda)(1) \tag{1.6.17}$$

is called the Nth Q-curvature polynomial of h. $Q_n^{res}(h; \lambda)$ is called the critical Q-curvature polynomial.

By $D_n^{res}(h; 0) = P_n(h)$ and $P_n(h)(1) = 0$, the critical Q-curvature polynomial has a vanishing constant term. Moreover, its linear term is given by

$$\dot{Q}_n^{res}(h; 0) = -(-1)^{\frac{n}{2}} \dot{D}_n^{res}(h; 0)(1) = Q_n(h)$$

by Theorem 1.5.4. Similarly, for $2N < n$,

$$Q_{2N}^{res}\left(h; -\frac{n}{2} + N\right) = -\left(\frac{n}{2} - N\right) Q_{2N}(h)$$

by

$$D_{2N}^{res}\left(h; -\frac{n}{2} + N\right) = P_{2N}(h)$$

(see the trivial case $j = 0$ of (1.6.6)). This motivates the terminology.

In more explicit terms, the Nth Q-curvature polynomial is defined by

$$\begin{aligned} Q_{2N}^{res}(h; \lambda) = -2^{2N} N! \left(\left(\lambda + \frac{n}{2} - 2N + 1\right) \cdots \left(\lambda + \frac{n}{2} - N\right) \right) \\ \times \left[\mathcal{T}_{2N}^*(h; \lambda + n - 2N)(v_0) + \cdots + \mathcal{T}_0^*(h; \lambda + n - 2N)(v_{2N}) \right]. \end{aligned} \tag{1.6.18}$$

In particular, the critical Q-curvature polynomial is defined by

$$\begin{aligned} &Q_n^{res}(h; \lambda) \\ &= -2^n \left(\frac{n}{2}\right)! \left(\left(\lambda - \frac{n}{2} + 1\right) \cdots \lambda \right) \left[\mathcal{T}_n^*(h; \lambda)(v_0) + \cdots + \mathcal{T}_0^*(h; \lambda)(v_n) \right]. \end{aligned} \tag{1.6.19}$$

The first two critical Q-polynomials can be determined by direct calculations.

Example 1.6.3. In dimension $n = 2$,

$$Q_2^{res}(\lambda) = \lambda Q_2.$$

Proof. By definition of $D_2^{res}(h;\lambda)$,

$$\begin{aligned} Q_2^{res}(h;\lambda) &= -4\lambda P_0^*(\lambda)(v_2) + P_2^*(\lambda)(v_0) \\ &= -4\lambda v_2 + (\Delta - \lambda \mathsf{J})(1) \\ &= 2\lambda \mathsf{J} - \lambda \mathsf{J} \\ &= \lambda Q_2(h) \end{aligned}$$

using $P_2(\lambda) = \Delta - \lambda \mathsf{J}$ and $v_2 = -\frac{1}{2}\mathsf{J}$. □

Example 1.6.4. In dimension $n = 4$,

$$Q_4^{res}(\lambda) = -\lambda(\lambda-1)Q_4 - \lambda^2 P_2(Q_2).$$

Proof. By definition of $D_4^{res}(h;\lambda)$,

$$-Q_4^{res}(\lambda) = 32(\lambda-1)\lambda P_0^*(v_4) + 8(-\lambda)P_2^*(\lambda)(v_2) + P_4^*(\lambda)(v_0).$$

The family $P_4(\lambda)$ is given in (1.6.4). A calculation yields

$$P_4^*(\lambda)(1) = -3\lambda\Delta\mathsf{J} + \lambda(\lambda+2)\mathsf{J}^2 + 2\lambda(\lambda-1)|\mathsf{P}|^2.$$

Hence using $P_2(\lambda) = \Delta - \lambda\mathsf{J}$, $v_2 = -\frac{1}{2}\mathsf{J}$ and $v_4 = \frac{1}{8}(\mathsf{J}^2 - |\mathsf{P}|^2)$, we find

$$Q_4^{res}(\lambda) = \lambda^2(2|\mathsf{P}|^2 - \mathsf{J}^2) + \lambda Q_4. \tag{1.6.20}$$

The latter result is equivalent to the assertion. □

Alternatively, these results can be derived from factorization identities for Q-polynomials which are consequences of the factorization identities for residue families. More precisely, Theorem 1.6.4 implies

Theorem 1.6.5. *For a locally conformally flat metric h, the Q-curvature polynomials satisfy the relations*

$$Q_{2N}^{res}\left(h; -\frac{n}{2} + 2N - j\right) = (-1)^j P_{2j}(h)\left(Q_{2N-2j}^{res}\left(h; -\frac{n}{2} + 2N - j\right)\right)$$

for $j = 1, \dots, N$.

Next, we observe that Theorem 1.5.3 implies that

$$e^{2N\varphi} Q_{2N}^{res}(e^{2\varphi}h; 0) = Q_{2N}^{res}(h; 0), \; \varphi \in C^\infty(M).$$

In other words, $Q_{2N}^{res}(h;0)$ is a scalar local conformal invariant of order $2N$. But for $2N < n$, there are no such non-trivial invariants for locally conformally flat metrics [FG07]. Thus, for locally conformally flat metrics h,

$$Q_{2j}^{res}(h;0) = 0 \quad \text{for} \quad j = 1, \ldots, \frac{n}{2}. \tag{1.6.21}$$

Note that in the critical case $2N = n$, this vanishing result follows from

$$Q_n^{res}(h;0) = -(-1)^{\frac{n}{2}} P_n(h)(1) = 0.$$

Now (1.6.21) implies that the polynomials

$$\mathcal{Q}_{2j}^{res}(h;\lambda) = Q_{2j}^{res}(h;\lambda)/\lambda, \; j = 1, \ldots, \frac{n}{2}$$

are well defined. These polynomials are still well-defined without assuming h to be locally conformally flat. This follows from the following result.

Theorem 1.6.6. *For even n, we have $Q_{2N}^{res}(h;0) = 0$ for all $1 \leq N \leq \frac{n}{2}$ and all metrics.*

Proof. We sketch the argument. First, assume that $n \geq 4N$, that is, $n - 2N \geq \frac{n}{2}$. For $j = 1, \ldots, N$, the poles of the family $\mathcal{T}_{2j}(h;\lambda)$ are in the set $\{\frac{n}{2} - 1, \ldots, \frac{n}{2} - j\}$. Thus, the assumption guarantees that the families $\mathcal{T}_2^*(h;\lambda), \ldots, \mathcal{T}_{2N}^*(h;\lambda)$ are regular in $n - 2N$. Now it follows from (1.6.18) that the assertion is equivalent to

$$\mathcal{T}_{2N}^*(h;n-2N)(v_0) + \cdots + \mathcal{T}_0^*(h;n-2N)(v_{2N}) = 0. \tag{1.6.22}$$

In turn, (1.6.22) is equivalent to

$$\int_M (v_0 \mathcal{T}_{2N}(h;n-2N)(f) + \cdots + v_{2N} \mathcal{T}_0(h;n-2N)(f)) \operatorname{vol}(h) = 0 \tag{1.6.23}$$

for all $f \in C_0^\infty(M)$. Let

$$u = r^{n-2N} f + r^{n-2N+2} \mathcal{T}_2(n-N)(f) + \cdots + r^n \mathcal{T}_{2N}(n-N)(f).$$

We choose small ε and δ so that $0 < \varepsilon < \delta$. Green's formula implies

$$-\int_{\varepsilon<r<\delta} \Delta_{g_+} u vol(g_+) = \left(\int_{r=\varepsilon} + \int_{r=\delta} \right) \frac{\partial u}{\partial \nu} r^{-n-1} v(r) \operatorname{vol}(h), \tag{1.6.24}$$

where ν denotes the inward normals; recall that $\operatorname{vol}(g_+) = r^{-n-1} v(r) dr \operatorname{vol}(h)$ with $v(r)$ as in (1.5.2). Both sides of (1.6.24) have asymptotic expansions for $\varepsilon \to 0$. We determine the respective coefficients of $\log \varepsilon$. By construction, we have

$$-\Delta_{g_+} u = 2N(n-2N)u + O(r^{n+2}).$$

It follows that the coefficient of $\log \varepsilon$ in the asymptotic expansion of the left-hand side of (1.6.24) is a non-trivial constant multiple of

$$\int_M (v_0 \mathcal{T}_{2N}(h;n-2N)(f) + \cdots + v_{2N} \mathcal{T}_0(h;n-2N)(f)) \operatorname{vol}(h).$$

On the other hand, the asymptotics of the right-hand side of (1.6.24) does not contain a $\log \varepsilon$ term. This proves (1.6.23). Thus, for fixed N, we have $\dot{Q}_{2N}^{res}(h;0) = 0$ for all $n \geq 4N$. The assertion for all $n \geq 2N$ follows by "analytic continuation" in the dimension n. ☐

The following result extends Example 1.6.3 and Example 1.6.4.

Theorem 1.6.7. *Let (M,h) be locally conformally flat of even dimension n. Then*

$$\begin{aligned} Q_n^{res}(h;\lambda) &= (-1)^{\frac{n}{2}-1} \lambda \prod_{k=1}^{\frac{n}{2}-1} \left(\frac{\lambda - \frac{n}{2} + k}{k} \right) Q_n(h) \\ &+ \lambda \sum_{j=1}^{\frac{n}{2}-1} (-1)^j \prod_{\substack{k=1 \\ k \neq j}}^{\frac{n}{2}} \left(\frac{\lambda - \frac{n}{2} + k}{k-j} \right) P_{2j}(h) \left(Q_{n-2j}^{res} \left(h; \frac{n}{2} - j \right) \right). \end{aligned} \quad (1.6.25)$$

Proof. For the moment being, denote the polynomial on the right-hand side of (1.6.25) by $R_n^{res}(\lambda)$. It is a polynomial of degree $\frac{n}{2}$. It suffices to prove that $R_n^{res}(h;\lambda)$ satisfies the relations

$$R_n^{res}\left(h; \frac{n}{2} - j\right) = (-1)^j P_{2j}(h) \left(Q_{2N-2j}^{res} \left(h; \frac{n}{2} - j \right) \right), \ j = 1, \dots, \frac{n}{2}$$

and

$$\dot{R}_n^{res}(h;0) = Q_n(h).$$

But it is straightforward to verify these assertions. ☐

Theorem 1.6.7 extends to subcritical Q-curvatures as follows.

Theorem 1.6.8. *Let (M,h) be locally conformally flat of even dimension n. Assume that $2N < n$. Then*

$$\begin{aligned} Q_{2N}^{res}(h;\lambda) &= (-1)^{N-1} \lambda \prod_{k=1}^{N-1} \left(\frac{\lambda + \frac{n}{2} - 2N + k}{k} \right) Q_{2N}(h) \\ &+ \lambda \sum_{j=1}^{N-1} (-1)^j \prod_{\substack{k=1 \\ k \neq j}}^{N} \left(\frac{\lambda + \frac{n}{2} - 2N + k}{k-j} \right) P_{2j}(h) \left(Q_{2N-2j}^{res} \left(h; -\frac{n}{2} + 2N - j \right) \right). \end{aligned}$$

In fact, for $2N < n$, the polynomial $Q_{2N}^{res}(h;\lambda)$ is characterized by the relations

$$Q_{2N}^{res}\left(h;-\frac{n}{2}+2N-j\right) = (-1)^j P_{2j}(h)\left(Q_{2N-2j}^{res}\left(h;-\frac{n}{2}+2N-j\right)\right)$$

for $j=1,\dots,N$, and the vanishing property

$$Q_{2N}^{res}(h;0) = 0.$$

Theorem 1.6.6 shows that the vanishing property (1.6.21) is actually satisfied for general metrics. This means that the formulas in Theorem 1.6.7 and Theorem 1.6.8 extend to all metrics.

Now combining Theorem 1.6.8 with (1.6.25), generates a formula for the critical Q-curvature polynomial $Q_n^{res}(\lambda)$ with coefficients given as linear combinations of compositions of GJMS-operators acting on lower-order Q-curvatures. The following result displays the coefficients of the respective leading powers in the critical polynomials $Q_{2N}^{res}(h;\lambda)$ for $N=1,\dots,4$.

Proposition 1.6.4. *The first four critical Q-curvature polynomials are of the form*

$$Q_2^{res}(\lambda) = \lambda Q_2, \tag{1.6.26}$$

$$Q_4^{res}(\lambda) = -\lambda^2\left(Q_4 + P_2(Q_2)\right) + \cdots, \tag{1.6.27}$$

$$2!Q_6^{res}(\lambda) = \lambda^3\left(Q_6 + 2P_2(Q_4) - 2P_4(Q_2) + 3P_2^2(Q_2)\right) + \cdots \tag{1.6.28}$$

and

$$\begin{aligned} 3!Q_8^{res}(\lambda) = -\lambda^4(&Q_8 + 3P_2(Q_6) + 3P_6(Q_2) - 9P_4(Q_4) - 8P_2P_4(Q_2) \\ &+ 12P_2^2(Q_4) - 12P_4P_2(Q_2) + 18P_2^3(Q_2)) + \cdots . \end{aligned} \tag{1.6.29}$$

(1.6.26) and (1.6.27) only restate Example 1.6.3 and Example 1.6.4. (1.6.28) and (1.6.29) hold true for general metrics. This rests on the fact that those factorization identities of residue families which involve a GJMS-operator of the boundary metric are valid in full generality. For the details we refer to [J09b].

Now recall that the critical residue families $D_n^{res}(h;\lambda)$ and the corresponding Q-curvature polynomials $Q_n^{res}(h;\lambda)$ are defined in terms of the holographic coefficients v_{2j} and the families $P_{2j}(h;\lambda) = \Delta^j + \cdots$ for $j = 0,\dots,\frac{n}{2}$. In particular, there are formulas for the leading coefficients of the polynomials $Q_n^{res}(\lambda)$ in terms of the holographic coefficients v_{2j} and the families $P_{2j}(\lambda)$. Comparing these formulas with those given in Proposition 1.6.4 yields interesting representations for the critical Q-curvatures Q_4, Q_6 and Q_8.

More precisely, we obtain the following results.

Proposition 1.6.5. *On any four-manifold,*

$$\begin{aligned} Q_4 + P_2(Q_2) &= -Q_2^2 + 16v_4 &(1.6.30)\\ &= 4(4v_4 - v_2^2). &(1.6.31)\end{aligned}$$

Proposition 1.6.6. *On any six-manifold,*

$$Q_6 + 2P_2(Q_4) - 2P_4(Q_2) + 3P_2^2(Q_2) = -6(Q_4 + P_2(Q_2))Q_2 - 2^6 3! v_6 \quad (1.6.32)$$
$$= -48(8v_6 - 4v_4 v_2 + v_2^3). \quad (1.6.33)$$

More details of the following proof can be found in Section 6.11 of [J09b].

Proof. We compare the coefficient of λ^3 in (1.6.28) with the coefficient of λ^3 in $2D_6^{res}(\lambda)(1)$, where

$$D_6^{res}(\lambda)(1) = 2^6 3!(-\lambda+2)(-\lambda+1)(-\lambda)[\mathcal{T}_6^*(\lambda)(v_0) + \mathcal{T}_4^*(\lambda)(v_2) + \mathcal{T}_2^*(\lambda)(v_4) + v_6],$$

or, equivalently,

$$D_6^{res}(\lambda)(1) = P_6^*(\lambda)(v_0) - 12\lambda P_4^*(\lambda)(v_2) + 2^4 3!\lambda(\lambda-1)P_2^*(\lambda)(v_4) - 2^6 3!\lambda(\lambda-1)(\lambda-2)v_6.$$

This yields the formula

$$Q_6 + 2P_2(Q_4) - 2P_4(Q_2) + 3P_2^2(Q_2) = -6\mathsf{J}^3 + 12|\mathsf{P}|^2\mathsf{J} - 2^6 3! v_6.$$

Now we recall the formulas $v_2 = -\frac{1}{2}\mathsf{J}$, $v_4 = \frac{1}{8}(\mathsf{J}^2 - |\mathsf{P}|^2)$,

$$P_2 = \Delta - 2\mathsf{J} \quad \text{and} \quad Q_4 = 3\mathsf{J}^2 - 2|\mathsf{P}|^2 - \Delta\mathsf{J}.$$

Using these results, it is straightforward to verify the assertions. □

It is also of interest to make (1.6.32) more explicit. A calculation shows that

$$-2P_2(Q_4) + 2P_4(Q_2) - 3P_2^2(Q_2) - 6(Q_4 + P_2(Q_2))Q_2 = \Delta^2\mathsf{J} + 4\Delta|\mathsf{P}|^2 + 8(\mathsf{P}, \operatorname{Hess}\mathsf{J}) - 8\mathsf{J}\Delta\mathsf{J}.$$

The latter sum can also be written in the form

$$-2P_2^0(Q_4) + 2P_4^0(Q_2) - 3P_2^0 P_2(Q_2),$$

where $(\cdot)^0$ denotes the non-constant part of the operator in brackets. Now (1.5.25) implies

$$-2^6 3! v_6 = 8(\mathsf{J}^3 - 3\mathsf{J}|\mathsf{P}|^2 + 2\operatorname{tr}\mathsf{P}^3) + 8(\mathcal{B}, \mathsf{P}).$$

Hence we have proved

Corollary 1.6.1. *The critical Q-curvature Q_6 is given by*

$$Q_6 = -2P_2^0(Q_4) + 2P_4^0(Q_2) - 3P_2^0 P_2(Q_2) - 2^6 3! v_6 \quad (1.6.34)$$

or, equivalently,

$$Q_6 = \Delta^2\mathsf{J} + 4\Delta|\mathsf{P}|^2 + 8(\mathsf{P}, \operatorname{Hess}\mathsf{J}) - 8\mathsf{J}\Delta\mathsf{J} + 8(\mathsf{J}^3 - 3\mathsf{J}|\mathsf{P}|^2 + 2\operatorname{tr}\mathsf{P}^3) + 8(\mathcal{B}, \mathsf{P}). \quad (1.6.35)$$

(1.6.35) coincides with the formula given in [GJ07]. It also coincides with the expression for the critical Q_6 derived in [GP03] (see also [Br05]). For a proof of the equivalence see Section 6.10 of [J09b].

Note that the first formula in Corollary 1.6.1 immediately confirms the relation

$$\int_M Q_6 \,\mathrm{vol} = -2^6 3! \int_M v_6 \,\mathrm{vol} \tag{1.6.36}$$

(see Corollary 1.5.2).

We continue with a description of analogous results for Q_8.

The following result is an analog of Proposition 1.6.6. Its proof in Section 6.13 of [J09b] makes use of a special case of a conjectural identity which generalizes Theorem 1.5.6. The assumption was removed in [J09d].

Proposition 1.6.7. *On any eight-manifold, the sum*

$$\begin{aligned} Q_8 &+ 3P_2(Q_6) + 3P_6(Q_2) - 9P_4(Q_4) \\ &- 8P_2P_4(Q_2) + 12P_2^2(Q_4) - 12P_4P_2(Q_2) + 18P_2^3(Q_2) \end{aligned} \tag{1.6.37}$$

equals

$$-12[Q_6+2P_2(Q_4)-2P_4(Q_2)+3P_2^2(Q_2)]Q_2-18[Q_4+P_2(Q_2)]^2+3!4!2^7 v_8. \tag{1.6.38}$$

Equivalently, (1.6.37) *equals*

$$288(64v_8 - 32v_6v_2 - 16v_4^2 + 24v_2^2v_4 - 5v_2^5). \tag{1.6.39}$$

Corollary 1.6.1 can be regarded as a description of cancellations in (1.6.32). There are similar cancellations of terms in (1.6.37) and (1.6.38). In fact, calculations yield the following result.

Corollary 1.6.2. *The critical Q-curvature Q_8 is given by the sum of*

$$\begin{aligned} &- 3P_2^0(Q_6) - 3P_6^0(Q_2) + 9P_4^0(Q_4) \\ &\qquad + 8P_2^0P_4(Q_2) - 12P_2^0P_2(Q_4) + 12P_4^0P_2(Q_2) - 18P_2^0P_2^2(Q_2), \end{aligned}$$

the divergence terms

$$\begin{aligned} 12\left[\Delta(Q_4)Q_2 - Q_4\Delta(Q_2)\right] &+ 18\left[\Delta^2(Q_2)Q_2 - \Delta(Q_2)\Delta(Q_2)\right] \\ &+ 64\left[\Delta(Q_2)Q_2^2 - Q_2\Delta(Q_2^2)\right] \end{aligned}$$

and the holographic coefficient

$$3!4!2^7 v_8.$$

Corollary 1.6.2 immediately confirms the relation

$$\int_M Q_8 \, \mathrm{vol} = 3!4!2^7 \int_M v_8 \, \mathrm{vol} \tag{1.6.40}$$

(see Corollary 1.5.2), and makes explicit all divergence terms.

Some comments are in order. The identities (1.6.30), (1.6.32) and (1.6.38) are recursive formulas for the critical Q_4, Q_6 and Q_8: they express critical Q-curvatures in terms of respective lower-order GJMS-operators acting on lower-order Q-curvatures. In addition, all formulas contain a contribution by a holographic anomaly v_n. These formulas resemble the holographic formula in the sense that both types of identities can be regarded as formulas for the differences

$$Q_n - 2^{n-1}(-1)^{\frac{n}{2}} \left(\frac{n}{2}\right)! \left(\frac{n}{2}-1\right)! v_n.$$

But both types of formulas express these differences in fundamentally different ways. While the holographic formula uses the holographic coefficients and the asymptotic expansion of harmonic functions for the Poincaré-Einstein metric, the formulas in Propositions 1.6.5–1.6.7 only use lower-order GJMS-operators and their constant terms, i.e., Q-curvatures.

In connection with the following comments, it is useful to recall that

$$Q_2 = -2v_2. \tag{1.6.41}$$

Now we observe that the first term on the right-hand side of (1.6.32) is composed of the quantities on the left-hand sides of (1.6.30) and (1.6.41). Similarly, the first two summands in (1.6.38) are composed of the left-hand sides of (1.6.30), (1.6.32) and (1.6.41).

The relations (1.6.31), (1.6.33) and (1.6.39) provide alternative formulas only in terms of holographic coefficients. The following result on power series implies that these formulas have a simple uniform formulation.

Lemma 1.6.1. *Let*

$$1 + w_2 r^2 + w_4 r^4 + w_6 r^6 + \cdots$$

be the Taylor series of the function $w(r) = \sqrt{v(r)}$ *with*

$$v(r) = 1 + v_2 r^2 + v_4 r^4 + v_6 r^6 + \cdots .$$

Then

$$\begin{aligned}
2w_2 &= v_2, \\
2w_4 &= \frac{1}{4}(4v_4 - v_2^2), \\
2w_6 &= \frac{1}{8}(8v_6 - 4v_4 v_2 + v_2^3), \\
2w_8 &= \frac{1}{64}(64v_8 - 32v_6 v_2 - 16v_4^2 + 24v_2^2 v_4 - 5v_2^4).
\end{aligned}$$

Thus, the right-hand sides of (1.6.41), (1.6.31) and (1.6.33) and the quantity (1.6.39) are given by

$$(-1)^N N!(N-1)!2^{2N} w_{2N} \tag{1.6.42}$$

for the appropriate values of N. These observations are special cases of the following conjecture [J09a].

Conjecture 1.6.2. *The leading coefficient of the critical Q-curvature polynomial $Q_n^{res}(h;\lambda)$ is given by*

$$-\left(\frac{n}{2}\right)!2^n w_n(h).$$

In [J09a], one also finds a discussion of an analogous conjecture for subcritical Q-curvature polynomials.

The above discussion shows that, in the formula

$$Q_6 = [-2P_2(Q_4) + 2P_4(Q_2) - 3P_2^2(Q_2)] - 6[Q_4 + P_2(Q_2)]Q_2 - 2^6 3! v_6,$$

the middle term

$$6[Q_4 + P_2(Q_2)]Q_2$$

is naturally linked *both* to the first and the last term. In fact, it naturally cancels certain contributions in the first term *and* contributes to w_6, when combined with v_6. Corollary 1.6.2 shows that for Q_8 the term (1.6.38) plays a similar double role.

We continue with a brief discussion of recursive formulas for some critical GJMS-operators. Such a formula for P_6 follows by combining the conformal variational formula

$$(d/dt)|_0 \left(e^{nt\varphi} Q_6(e^{2t\varphi} h)\right) = -P_6(h)(\varphi)$$

with the recursive formula

$$Q_6 = -2P_2(Q_4) + 2P_4(Q_2) - 3P_2^2(Q_2) - 6(Q_4 + P_2(Q_2))Q_2 - 2^6 3! v_6$$

for the critical Q_6 (see (1.6.32)). We also recall that v_6 is given by (1.5.25). This leads to the following result.

Theorem 1.6.9. *The critical GJMS-operator P_6 coincides with the operator*

$$(2P_2P_4 + 2P_4P_2 - 3P_2^3)^0 - \delta((48\mathsf{P}^2 + 8\mathcal{B})\#d). \tag{1.6.43}$$

We recall that the superscript $(\cdot)^0$ denotes the non-constant part of the operator in brackets. A detailed proof of Theorem 1.6.9 is given in Section 6.12 of [J09b].

The GJMS-operator P_6 exists also in all odd dimensions $n \geq 3$ and in all even dimensions $n \geq 6$. These cases are covered by the following generalization of Theorem 1.6.9.

Theorem 1.6.10. *For odd $n \geq 3$ and even $n \geq 6$, the GJMS-operator P_6 on manifolds of dimension n is given by*

$$(2P_2P_4 + 2P_4P_2 - 3P_2^3)^0 - 48\delta(\mathsf{P}^2\#d) - \frac{16}{n-4}\delta(\mathcal{B}\#d) - \left(\frac{n}{2}-3\right) Q_6 \qquad (1.6.44)$$

with

$$Q_6 = \left[-2P_2(Q_4) + 2P_4(Q_2) - 3P_2^2(Q_2)\right] - 6\left[Q_4 + P_2(Q_2)\right] Q_2 - 2^6 3! v_6. \quad (1.6.45)$$

Here

$$v_6 = -\frac{1}{8}\operatorname{tr}\wedge^3(\mathsf{P}) - \frac{1}{24(n-4)}(\mathcal{B}, \mathsf{P}). \qquad (1.6.46)$$

From the perspective of these results for P_6, it is natural to reformulate the usual formula for the Paneitz operator P_4 as follows.

Theorem 1.6.11. *For odd $n \geq 3$ and even $n \geq 4$, the Paneitz operator P_4 on manifolds of dimension n is given by*

$$(P_2^2)^0 - 4\delta(\mathsf{P}\#d) + \left(\frac{n}{2}-2\right) Q_4 \qquad (1.6.47)$$

with

$$Q_4 = -P_2(Q_2) - Q_2^2 + 16v_4. \qquad (1.6.48)$$

We stress the main features of the formulas (1.6.47) and (1.6.44) for P_4 and P_6. Both formulas express the respective GJMS-operator as a sum of three terms. The first terms are the non-constant parts of linear combinations of products of lower-order GJMS-operators. In that sense, the formulas are recursive. The second parts are *second-order* operators which are determined by a power of the Schouten tensor P and the Bach tensor $\mathcal{B}$. The appropriate interpretation of the quotient $\frac{\mathcal{B}}{n-4}$ is as an *extended obstruction tensor* in the sense of [G09]. Note that the right-hand side of (1.6.44) is not well defined for $n = 4$. This reflects the fact that there does not exist a conformally covariant cube of the Laplacian on general four-manifolds [G92]. Of course, the third terms in both formulas are given by Q-curvatures. Finally, we emphasize that the first terms in both formulas are relatives of the first terms in the corresponding recursive formula for Q-curvatures. In fact, the former arise by substituting Q-curvatures in the latter by the corresponding GJMS-operator (up to a sign).

Theorem 1.6.10 was first proved in [J09b]. In [J09a], the result was embedded into a much wider framework, which yields an analogous formula for a conformally covariant fourth power of the Laplacian, as well as conjectural formulas for all higher-order powers (for locally conformally flat metrics).

We close this section with a discussion of some results concerning renormalized volumes. We recall from Section 1.5 that (for even n) the quantity v_n is the conformal anomaly of the renormalized volume $V(g_+, h)$ of a Poincaré-Einstein metric g_+ with conformal infinity $[h]$, i.e.,

$$(d/dt)|_0 \left(V(e^{2t\varphi}h)\right) = \int_M \varphi v_n(h) \operatorname{vol}(h). \tag{1.6.49}$$

The full anomaly, i.e., the integrand $\mathcal{A}_n(\varphi)$ in

$$V(e^{2\varphi}h) - V(h) = \int_M \mathcal{A}_n(\varphi) \operatorname{vol}(h),$$

is determined by v_n through

$$\int_M \mathcal{A}_n(\varphi) \operatorname{vol}(h) = \int_0^1 (d/dt)(V(h_t))dt = \int_0^1 \left(\int_M \varphi v_n(h_t) \operatorname{vol}(h_t)\right) dt.$$

$\mathcal{A}$ is a natural *non-linear* differential operator.

Now we shall show how one can use the relations between v_2 and Q_2, v_4 and Q_4, and v_6 and Q_6 to derive some compact formulas for the respective full anomalies.

We start with a closed surface M. Using $v_2 = -\frac{1}{2}K = -\frac{1}{2}Q_2$ and the transformation law (1.2.12), we find

$$\begin{aligned} V(e^{2\varphi}h) - V(h) &= \int_0^1 \left(\int_M \varphi v_2(h_t) \operatorname{vol}(h_t)\right) dt \\ &= -\frac{1}{2}\int_0^1 \left(\int_M \varphi Q_2(h_t) \operatorname{vol}(h_t)\right) dt \\ &= -\frac{1}{2}\left(\int_M \varphi Q_2(h) \operatorname{vol}(h) - \frac{1}{2}\int_M \varphi P_2(h)(\varphi) \operatorname{vol}(h)\right) \\ &= -\frac{1}{4}\int_M \varphi \left[Q_2(e^{2\varphi}h) \operatorname{vol}(e^{2\varphi}h) + Q_2(h) \operatorname{vol}(h)\right]. \end{aligned}$$

This transformation law should be compared with Polyakov's formula (1.2.37) (combined with (1.2.38)) for the determinant of the Laplacian.

We continue with a discussion of the renormalized volume of g_+ if the boundary is a closed four-manifold. It will be useful to observe that the relation

$$Q_4 + P_2(Q_2) = -Q_2^2 + 16v_4$$

(see (1.6.30)) holds true not only in dimension $n = 4$ but in *all* dimensions $n \geq 3$. This can be verified by a direct calculation. We write the relation in the form

$$Q_4 - 16v_4 = \frac{n-4}{2}\mathsf{J}^2 - \Delta\mathsf{J}$$

and integrate over the closed manifold M^n. Then

$$\int_M (Q_4 - 16v_4) \operatorname{vol} = \frac{n-4}{2}\int_M \mathsf{J}^2 \operatorname{vol}$$

for any metric h. Now we determine the infinitesimal conformal variation of this identity. It will be convenient to use the abbreviations $\hat{h} = e^{2\varphi}h$ and $\hat{h}_t = e^{2t\varphi}h$. In view of

$$(d/dt)|_0 \left(\int_M Q_4(\hat{h}_t) \operatorname{vol}(\hat{h}_t) \right) = (n-4) \int_M \varphi Q_4(h) \operatorname{vol}(h)$$

(see [Br93]) and

$$(d/dt)|_0 \left(\int_M v_4(\hat{h}_t) \operatorname{vol}(\hat{h}_t) \right) = (n-4) \int_M \varphi v_4(h) \operatorname{vol}(h)$$

(see the proof of Theorem 1.2.9) we find

$$(n-4) \int_M \varphi \left(Q_4(h) - 16 v_4(h)\right) \operatorname{vol}(h) = \frac{n-4}{2} (d/dt)|_0 \left(\int_M \mathsf{J}^2(\hat{h}_t) \operatorname{vol}(h_t) \right).$$

We divide the latter relation by $n-4$ and set $n=4$. This yields the relation

$$\int_{M^4} \varphi \left(Q_4(h) - 16 v_4(h)\right) \operatorname{vol}(h) = \frac{1}{2} (d/dt)|_0 \left(\int_{M^4} \mathsf{J}^2(\hat{h}_t) \operatorname{vol}(\hat{h}_t) \right) \qquad (1.6.50)$$

on any four-manifold. (1.6.50) implies

$$V(\hat{h}) - V(h) = \frac{1}{16} \int_0^1 \left(\int_M \varphi Q_4(h_t) \operatorname{vol}(h_t) \right) dt - \frac{1}{32} \int_0^1 (d/ds)|_0 \left(\int_M \mathsf{J}^2(\hat{h}_{t+s}) \operatorname{vol}(\hat{h}_{t+s}) \right) dt.$$

Now using the transformation law (1.2.23), the latter result simplifies to

$$V(\hat{h}) - V(h) = \frac{1}{32} \int_{M^4} \varphi \left[Q_4(\hat{h}) \operatorname{vol}(\hat{h}) + Q_4(h) \operatorname{vol}(h) \right] - \frac{1}{32} \int_{M^4} \left[\mathsf{J}^2(\hat{h}) \operatorname{vol}(\hat{h}) - \mathsf{J}^2(h) \operatorname{vol}(h) \right]. \qquad (1.6.51)$$

Again, the structure of this formula resembles the structure of Polyakov's formula in Theorem 1.2.7. A direct calculation shows that (1.6.51) coincides with a formula in [G00], up to a correcting factor.

Note that we derived the identity (1.6.50) in dimension $n = 4$ through consideration of related identities in dimension $\neq 4$. This method mimics a technique which was used by Branson [Br93], [Br95] to derive conformal variational formulas for determinants. The power of the method is even more apparent in the following case.

The recursive formula (1.6.32) extends to general dimensions $n \geq 5$ (see [J09b]) and a calculation yields

$$V(\hat{h}) - V(h) = -\frac{1}{2^6 3!2!}$$
$$\times \left(\int_{M^6} \varphi \left[Q_6(\hat{h}) \operatorname{vol}(\hat{h}) + Q_6(h) \operatorname{vol}(h) \right] - \int_{M^6} \left[\mathcal{R}_6(\hat{h}) \operatorname{vol}(\hat{h}) - \mathcal{R}_6(h) \operatorname{vol}(h) \right] \right)$$

with

$$\mathcal{R}_6 = Q_2(4Q_4 + 3P_2(Q_2)).$$

The proof of this formula makes use of the relations

$$(d/dt)|_0 \left(\int_M Q_6(\hat{h}_t) \operatorname{vol}(\hat{h}_t) \right) = (n-6) \int_M \varphi Q_6(h) \operatorname{vol}(h)$$

see [Br93]) and

$$(d/dt)|_0 \left(\int_M v_6(\hat{h}_t) \operatorname{vol}(\hat{h}_t) \right) = (n-6) \int_M \varphi v_6(h) \operatorname{vol}(h). \tag{1.6.52}$$

For proofs of (1.6.52) we refer to [CF08] and [G09]. In fact, an analog of (1.6.52) for v_{2N}, $2N < n$ leads to a natural analog of Theorem 1.2.9 for the functional

$$\int_M v_{2N} \operatorname{vol} \Big/ \left(\int_M \operatorname{vol} \right)^{\frac{n-2N}{n}}.$$

The above formulas should play a role in connection with the problem to understand the extremal properties of the renormalized volume.

Chapter 2

Conformal holonomy

2.1 Cartan connections and holonomy groups

In this section we give a short introduction to Cartan connections and define their holonomy groups. In particular, we explain the relation to holonomy groups of principal fibre bundle connections and to holonomy groups of covariant derivatives in associated vector bundles. Details can be found in [KN63], [Sh97] and [Ba09].

Let M be a smooth manifold and let $\mathcal{L}(M,x)$ denote the set of all piecewise smooth loops in a point $x \in M$. Usually, holonomy groups are defined by parallel displacements along loops in x with respect to a connection form on a principal fibre bundle or with respect to a covariant derivative in a vector bundle over M given by the geometric structure in question. Let us briefly recall this in order to fix some notation.

At first, we consider a vector bundle E over M with covariant derivative ∇^E. For any piecewise smooth curve $\gamma : [\alpha,\beta] \subset \mathbb{R} \to M$ joining $x = \gamma(\alpha)$ with $y = \gamma(\beta)$, the covariant derivative gives rise to a parallel displacement between the fibres of E over x and over y:

$$\begin{array}{rccc} \mathcal{P}_\gamma^{\nabla^E} : & E_x & \longrightarrow & E_y \\ & e & \longmapsto & \varphi_e(\beta) \end{array}, \tag{2.1.1}$$

where $\varphi_e : [\alpha,\beta] \to E$ is the unique parallel section in E along γ defined by

$$\frac{\nabla^E \varphi_e}{dt} = 0 \qquad \text{and} \qquad \varphi_e(\alpha) = e.$$

The *holonomy group of* (E,∇^E) *with respect to the base point* x is the Lie group of all parallel displacements along loops in x:

$$\mathrm{Hol}_x(E,\nabla^E) := \{\mathcal{P}_\gamma^{\nabla^E} \mid \gamma \in \mathcal{L}(M,x)\} \subset \mathrm{GL}(E_x).$$

Next, let $(P, \pi, M; B)$ be a principal fibre bundle over M with structure group B. We consider a connection form A on P,, i.e., a 1-form $A \in \Omega^1(P, \mathfrak{b})$ with values in the Lie algebra $\mathfrak{b}$ of B, which satisfies the invariance conditions

1. $R_b^* A = \mathrm{Ad}(b^{-1}) \circ A$ for all $b \in B$ and
2. $A(\widetilde{X}) = X$ for all $X \in \mathfrak{b}$,

where R_b denotes the right action of $b \in B$ on the total space P of the principal fibre bundle and $\widetilde{X}$ is the fundamental vector field of the B-action on P defined by the element $X \in \mathfrak{b}$,

$$\widetilde{X}(u) := \frac{d}{dt}\Big(u \cdot \exp(tX)\Big)_{|t=0}.$$

Let $u \in P_x := \pi^{-1}(x) \subset P$ be a point in the fibre of P over $x \in M$. The subspace

$$Tv_u P := T_u P_x = \{\widetilde{X}(u) \mid X \in \mathfrak{b}\} \subset T_u P$$

is called the *vertical tangent space* in $u \in P$. A linear subspace $H \subset T_u P$ is called *horizontal*, if $T_u P = H \oplus Tv_u P$. The connection form A defines a right invariant horizontal distribution $\mathrm{Th}^A P$ on P by

$$\mathrm{Th}^A : u \in P \longrightarrow \mathrm{Th}_u^A P := \ker A_u \subset T_u P.$$

On the other hand, any right invariant horizontal distribution $\mathrm{Th} : u \in P \mapsto \mathrm{Th}_u P \subset T_u P$ gives us a connection form $A \in \Omega^1(P, \mathfrak{b})$ by setting

$$A_u(Y \oplus \widetilde{X}(u)) := X \in \mathfrak{b},$$

where $Y \in \mathrm{Th}_u P$. A piecewise smooth curve $\gamma : [\alpha, \beta] \to M$ with initial point $x = \gamma(\alpha)$ admits a unique horizontal lift $\gamma_u^* : [\alpha, \beta] \to P$ with fixed initial point $u \in P_x$, defined by the conditions

$$\dot{\gamma}^*(t) \in \mathrm{Th}^A_{\gamma^*(t)} P\,, \quad \pi(\gamma^*(t)) = \gamma(t) \quad \text{and} \quad \gamma^*(\alpha) = u.$$

Thus, for a curve $\gamma : [\alpha, \beta] \to M$, the connection form A on P defines a parallel displacement between the fibres P_x and P_y of P by

$$\mathcal{P}_\gamma^A : \begin{array}{ccc} P_x & \longmapsto & P_y \\ u & \longmapsto & \gamma_u^*(\beta) \end{array}. \tag{2.1.2}$$

If γ is a loop in x, then the points u and $\mathcal{P}_\gamma^A(u)$ lie in the same fibre P_x. Hence there is a unique element $\mathrm{hol}_u^A(\gamma) \in B$ with

$$u = \mathcal{P}_\gamma^A(u) \cdot \mathrm{hol}_u^A(\gamma),$$

called the holonomy of the loop γ with respect to A and u. The *holonomy group of* (P, A) *with respect to the base point* u is the Lie group of all of these holonomies,

$$\mathrm{Hol}_u(P, A) := \{\mathrm{hol}_u^A(\gamma) \mid \gamma \in \mathcal{L}(M, x)\} \subset B.$$

Now, let $\rho : B \to GL(V)$ be a representation of the Lie group B over a vector space V. There is a standard way to associate a vector bundle E over M to the principal fibre bundle P by means of ρ. The total space E is defined to be the orbit space of the right action of B on $P \times V$ given by

$$(u, v) \cdot b := (u \cdot b,\ \rho(b^{-1})v)\ , \qquad (u, v) \in P \times V\ ,\ b \in B.$$

We denote this orbit space by $E := P \times_B V := (P \times V)/B$ and the elements of E by $[u, v] := \{(u \cdot b, \rho(b^{-1})v) \mid b \in B\}$. Obviously, E is a vector bundle over M with fibre type V and projection $\pi_E([u, v]) := \pi(u)$. Any point u in the fibre P_x of P over $x \in M$ gives rise to a linear isomorphism

$$[u] : v \in V \longrightarrow [u, v] \in E_x := \pi_E^{-1}(x) \tag{2.1.3}$$

between the fibre type V and the fibre E_x of E. A smooth section ϕ in the vector bundle E can locally be represented in the form $\phi_{|U} = [s, v]$, where $s : U \subset M \to P$ is a smooth local section in the principal fibre bundle P and $v \in C^\infty(U, V)$ is a smooth function on U with values in the vector space V. Any connection form $A \in \Omega^1(P, \mathfrak{b})$ on P induces a covariant derivative $\nabla^A : \Gamma(E) \to \Gamma(T^*M \otimes E)$ by

$$\nabla_X^A \phi\,_{|U} := [s, X(v) + \rho_*\big(A(ds(X))\big)v]\ , \quad \text{where } \phi_{|U} = [s, v]. \tag{2.1.4}$$

The formulas (2.1.1), (2.1.2) and (2.1.4) show that the holonomy groups of (E, ∇^A) and (P, A) are related by

$$\mathrm{Hol}_x(E, \nabla^A) = [u] \circ \rho\big(\mathrm{Hol}_u(P, A)\big) \circ [u]^{-1}. \tag{2.1.5}$$

Hence, if the representation ρ is faithful, the holonomy groups $\mathrm{Hol}_x(E, \nabla^A)$ and $\mathrm{Hol}_u(P, A)$ are isomorphic.

Now, let us come to the objects of our interest, to Cartan connections and their holonomy groups. In order to define a Cartan connection on the principal fibre bundle P, we consider in addition a Lie group G that contains the structure group B of P as a closed subgroup.

Definition 2.1.1. A Cartan connection of type G/B on the B-principal fibre bundle P is a 1-form $\omega \in \Omega^1(P, \mathfrak{g})$ on P with values in the Lie algebra $\mathfrak{g}$ (!!) such that

1. $R_b^*\omega = \mathrm{Ad}(b^{-1}) \circ \omega$ for all $b \in B \subset G$,
2. $\omega(\widetilde{X}) = X$ for all $X \in \mathfrak{b} \subset \mathfrak{g}$ and
3. ω is a global parallelism, i.e., $\omega_u : T_uP \to \mathfrak{g}$ is an isomorphism for all $u \in P$.

Note that the third condition implies that the base manifold M has the same dimension as the homogeneous space G/B. Furthermore, any vector $Z \in \mathfrak{g}$ induces a global nowhere-vanishing vector field Z_ω on P by

$$Z_\omega(u) := (\omega_u)^{-1}(Z).$$

Hence, the manifold P is parallelizable, i.e., it has a global basis field. A Cartan connection is called *complete* if all these vector fields Z_ω are complete (meaning that the integral curves of Z_ω are defined for all parameters $t \in \mathbb{R}$).

Let us consider two examples of Cartan connections:

Example 2.1.1. Let G be a Lie group and $\omega_G \in \Omega^1(G, \mathfrak{g})$ its Maurer-Cartan form:

$$\omega_G : X \in T_gG \longmapsto dL_{g^{-1}}(X) \in T_eG = \mathfrak{g}.$$

If $B \subset G$ is a closed subgroup, then the projection $\pi : G \longrightarrow G/B$ gives rise to a principal fibre bundle with structure group B over the homogeneous space $M = G/B$. The Maurer-Cartan form ω_G is a complete Cartan connection of type G/B on this homogeneous bundle.

Example 2.1.2. Let M^n be a smooth manifold, and let P be the frame bundle of M. Then the tangent bundle of M is given by $TM = P \times_{\mathrm{GL}(n)} \mathbb{R}^n$. Furthermore, let $A \in \Omega^1(P, \mathfrak{gl}(n))$ be an arbitrary connection form on P, and let $\theta \in \Omega^1(P, \mathbb{R}^n)$ be the displacement form of P,

$$\theta_u(X) := [u]^{-1} d\pi_u(X) \qquad u \in P,\ X \in T_uP. \tag{2.1.6}$$

Then the 1-form $\omega := A + \theta \in \Omega^1(P, \mathfrak{a}(n))$ is a Cartan connection of type $\mathrm{A}(n)/\mathrm{GL}(n)$, where $\mathrm{A}(n) := \mathrm{GL}(n) \ltimes \mathbb{R}^n$ is the affine group and $\mathfrak{a}(n)$ its Lie algebra.

Although, contrary to usual connections, Cartan connections do not allow one to distinguish a right invariant horizontal distribution on P, one can define the notion of a holonomy group also for Cartan connections using the development of the connection form along curves. By the *development of a* 1*-form along a curve* we understand the following: Let N be a manifold, and let $\sigma \in \Omega^1(N; \mathfrak{g})$ be a 1-form with values in the Lie algebra $\mathfrak{g}$ of a Lie group G. For any piecewise smooth curve $\lambda : [\alpha, \beta] \to N$ there is a unique curve $\lambda_{[\sigma]} : [\alpha, \beta] \to G$ starting in the unit element $e \in G$ such that

$$\lambda'_{[\sigma]}(t) = dL_{\lambda_{[\sigma]}(t)}\, \omega\big(\lambda'(t)\big), \tag{2.1.7}$$

where L_g denotes the left multiplication with the element $g \in G$. We call $\lambda_{[\sigma]} : [\alpha, \beta] \to G$ the *development of* σ *along* λ and its endpoint

$$\mathrm{hol}^\sigma(\lambda) := \lambda_{[\sigma]}(\beta) \tag{2.1.8}$$

the *holonomy of* λ *with respect to* σ.

Example 2.1.3. Let $G = (\mathbb{R}^n, +)$ be the additive group. Then the development of a 1-form $\sigma \in \Omega^1(N, \mathbb{R}^n)$ along a curve λ in N is the usual curve integral of σ along the curve λ,

$$\lambda_{[\sigma]}(t) := \int_\alpha^t \sigma(\lambda'(s))\, ds\,, \qquad \text{and} \qquad \mathrm{hol}^\sigma(\lambda) = \int_\lambda \sigma.$$

Example 2.1.4. Let ω_G be the Maurer-Cartan form of a Lie group G and λ a curve in G from g_0 to g_1. Then the development of ω_G along λ is the curve $\lambda_{[\omega_G]}$ given by

$$\lambda_{[\omega_G]}(t) = g_0^{-1} \cdot \lambda(t).$$

The holonomy is $\mathrm{hol}^{\omega_G}(\lambda) = g_0^{-1} \cdot g_1$. In particular, the holonomy of any closed curve λ with respect to ω_G is trivial.

Now, let us come back to the B-principal fibre bundle P with a given Cartan connection $\omega \in \Omega^1(P, \mathfrak{g})$. We fix a point u in the fibre P_x over $x \in M$ and denote by $\pi(\mathcal{L}(P,u))$ the set of all loops in $\mathcal{L}(M,x)$ which admit a closed lift in $\mathcal{L}(P,u)$. One observes that $\mathrm{hol}^\omega(\delta_1) = \mathrm{hol}^\omega(\delta_2)$ for loops $\delta_1, \delta_2 \in \mathcal{L}(P,u)$ with the same projection $\pi \circ \delta_1 = \pi \circ \delta_2 \in \mathcal{L}(M,x)$. Thus, we can define the *holonomy of a loop* $\gamma \in \pi(\mathcal{L}(P,u)) \subset \mathcal{L}(M,x)$ *with respect to* ω *and* u by

$$\mathrm{hol}_u^\omega(\gamma) := \mathrm{hol}^\omega(\widetilde{\gamma}) \in G\,, \qquad \text{where } \widetilde{\gamma} \in \mathcal{L}(P,u) \text{ and } \pi \circ \widetilde{\gamma} = \gamma. \tag{2.1.9}$$

The *holonomy group of the Cartan connection* ω *with respect to the base point* u is the group

$$\mathrm{Hol}_u(P,\omega) := \{\, \mathrm{hol}_u^\omega(\gamma) \mid \gamma \in \pi(\mathcal{L}(P,u)) \subset \mathcal{L}(M,x)\} \subset G.$$

$\mathrm{Hol}_u(P,\omega)$ is an immersed Lie subgroup of G with the *reduced holonomy group*

$$\mathrm{Hol}_u^0(P,\omega) := \{\, \mathrm{hol}_u^\omega(\gamma) \mid \gamma \in \mathcal{L}(M,x) \text{ null homotopic}\,\}$$

as connected component of the unit element. Holonomy groups for different reference points u are conjugated in G. Therefore, we often omit the base point and understand the holonomy group as a class of conjugated subgroups in the group G in question.

Example 2.1.5. Consider the Maurer-Cartan form ω_G as a Cartan connection on the homogeneous principal fibre bundle $(G, \pi, G/B; B)$. Then, by Example 2.1.4, the holonomy group is trivial: $\mathrm{Hol}_u(G, \omega_G) = \{e\}$.

Now, let $\rho : G \to GL(V)$ be a representation of the larger group G on a vector space V. Of course, by restriction, this gives a representation of B on V and an associated vector bundle

$$E := P \times_B V.$$

Bundles of this form (defined using a representation of the larger group G) are called *tractor bundles* and play a crucial role in Cartan geometry – since, contrary to the case of arbitrary associated vector bundles, on a tractor bundle there is a covariant derivative associated to the Cartan connection ω. To see this, we extend the B-principal fibre bundle P to the G-principal fibre bundle $\overline{P} := P \times_B G$. There is an easy way to extend the Cartan connection ω on P to a principal bundle connection $\overline{\omega}$ on $\overline{P}$: First we use the Cartan connection ω on P and the Maurer-Cartan form ω_G of G to define a 1-form $\hat{\omega} \in \Omega^1(P \times G, \mathfrak{g})$ by

$$\hat{\omega}_{(u,g)} := \mathrm{Ad}(g^{-1}) \circ (\pi_P^* \omega)_{(u,g)} + (\pi_G^* \omega_G)_{(u,g)}, \tag{2.1.10}$$

where π_P and π_G are the projections from $P \times G$ onto P and G, respectively. The 1-form $\hat{\omega}$ is invariant under the B-action on $P \times G$ and therefore projects to a 1-form $\overline{\omega} \in \Omega^1(\overline{P}, \mathfrak{g})$. A direct calculation shows that $\overline{\omega}$ is indeed a usual connection form on the G-principal fibre bundle $\overline{P}$. For the canonical embedding $\iota : u \in P \mapsto [u, e] \in \overline{P}$ obviously holds $\iota^* \overline{\omega} = \omega$. Since $\rho : G \to GL(V)$ is a G-representation, we have the vector bundle isomorphism

$$E = P \times_B V \simeq \overline{P} \times_G V.$$

Hence, any Cartan connection $\omega \in \Omega^1(P, \mathfrak{g})$ defines a covariant derivative ∇^ω on E via its extension to the principal fibre bundle connection $\overline{\omega}$ on $\overline{P}$, namely $\nabla^\omega := \nabla^{\overline{\omega}}$. According to formula (2.1.4), ∇^ω is given by

$$(\nabla^\omega_X \phi)_{|U} = [s, X(v) + \rho_*(\omega(ds(X)))v], \tag{2.1.11}$$

where $\phi_{|U} = [s, v] \in \Gamma(E_{|U})$ for a local section $s : U \to P$ and a smooth function $v : U \to V$. By definition, ∇^ω is a metric covariant derivative with respect to any metric $\langle \cdot, \cdot \rangle$ on the bundle E, which is induced by a G-invariant inner product $\langle \cdot, \cdot \rangle_V$ on V, meaning that

$$X(\langle \varphi, \psi \rangle) = \langle \nabla^\omega_X \varphi, \psi \rangle + \langle \varphi, \nabla^\omega_X \psi \rangle$$

for all sections $\varphi, \psi \in \Gamma(E)$ and vector fields $X \in \mathfrak{X}(M)$.

Problem 2.1.1. *Show that there is a 1-1-correspondence between the set of Cartan connections on P and the set of connection forms $\overline{\omega}$ on $\overline{P}$ with the additional condition* $\ker \overline{\omega} \cap d\iota(TP) = \{0\}$.

Example 2.1.6. We consider the situation of Example 2. Let P be the frame bundle over M^n with Cartan connection $\omega = A + \theta$, where A is a connection form and θ the displacement form on P. We realize the affine group $\mathrm{A}(n)$ as a subgroup of the linear group $\mathrm{GL}(n+1)$ by

$$(A, x) \in \mathrm{A}(n) = \mathrm{GL}(n) \ltimes \mathbb{R}^n \mapsto \begin{pmatrix} A & x \\ 0 & 1 \end{pmatrix} \in \mathrm{GL}(n+1)$$

and consider its matrix representation ρ on $\mathbb{R}^{n+1}$. The restriction of this representation to $\mathrm{GL}(n)$ decomposes into the sum $\mathbb{R}^n \oplus \mathbb{R}$, where the first factor is the matrix representation of $\mathrm{GL}(n)$ on $\mathbb{R}^n$ and the second the trivial representation. Hence, the tractor bundle $\mathcal{T}$ associated to P using ρ is

$$\mathcal{T} = P \times_{\mathrm{GL}(n)} \mathbb{R}^{n+1} \approx TM \oplus \underline{\mathbb{R}},$$

where $\underline{\mathbb{R}}$ denotes the trivial line bundle on M. For the tractor connection ∇^ω on $\mathcal{T}$ induced by ω we obtain from (2.1.4), (2.1.6) and (2.1.11)

$$\nabla^\omega_X \begin{pmatrix} Z \\ f \end{pmatrix} = \begin{pmatrix} \nabla^A_X Z + fX \\ X(f) \end{pmatrix},$$

where f is a function and Z a vector field on M, and ∇^A is the covariant derivative on the tangent bundle induced by A.

Next, let us discuss the relation between the holonomy groups of ω, $\overline{\omega}$ and ∇^ω which we defined above[1].

Proposition 2.1.1.

1. *The connected components of the holonomy groups coincide;*

$$\mathrm{Hol}^0_u(P,\omega) = \mathrm{Hol}^0_{[u,e]}(\overline{P},\overline{\omega}). \tag{2.1.12}$$

2. *If the structure group B of P is connected or if the base space M of P is simply connected, then*

$$\mathrm{Hol}_u(P,\omega) = \mathrm{Hol}_{[u,e]}(\overline{P},\overline{\omega}) \ \subset \ G, \tag{2.1.13}$$

$$\rho(\mathrm{Hol}_u(P,\omega)) = [u]^{-1} \circ \mathrm{Hol}_x(E,\nabla^\omega) \circ [u] \ \subset \ \mathrm{GL}(V). \tag{2.1.14}$$

Proof. Let $\gamma \in \mathcal{L}(M,x)$ be a loop in x which has a closed lift $\widetilde{\gamma} \in \mathcal{L}(P,u)$. First we will show that the holonomy of such curves γ with respect to the Cartan connection $\omega \in \Omega^1(P,\mathfrak{g})$ and the point $u \in P$ coincide with the holonomy of γ with respect to the connection form $\overline{\omega} \in \Omega^1(\overline{P},\mathfrak{g})$ and the point $\overline{u} := [u,e] \in \overline{P}$.

Let $\gamma^*_{\overline{u}} : [\alpha,\beta] \to \overline{P} := P \times_B G$ be the $\overline{\omega}$-horizontal lift of γ in $\overline{P}$ with initial point $\overline{u}$. By definition of $\overline{P}$ we can represent the curve $\gamma^*_{\overline{u}}$ in the form $\gamma^*_{\overline{u}} = [\widetilde{\gamma}, g]$, where $g : [\alpha,\beta] \to G$ and $g(\alpha) = e$. Then,

$$\gamma^*_{\overline{u}}(\beta) = [\widetilde{\gamma}(\beta), g(\beta)] = [u, g(\beta)] = [u,e]\cdot g(\beta) = \overline{u}\cdot g(\beta).$$

Hence, by definition, $\mathrm{hol}^{\overline{\omega}}_{\overline{u}}(\gamma) = g(\beta)^{-1}$. Since $\gamma^*_{\overline{u}}$ is $\overline{\omega}$-horizontal, we obtain from the definition of $\overline{\omega}$ and formula (2.1.10)

$$0 = \overline{\omega}_{\gamma^*_{\overline{u}}(t)}\big(\dot{\gamma}^*_{\overline{u}}(t)\big) = \hat{\omega}_{(\widetilde{\gamma}(t),g(t))}\big(\dot{\widetilde{\gamma}}(t), \dot{g}(t)\big) = \mathrm{Ad}\,\big(g^{-1}(t)\big)\,\omega\big(\dot{\widetilde{\gamma}}(t)\big) + \omega_G\big(\dot{g}(t)\big).$$

[1]There are several authors in the literature who define the holonomy group of a Cartan connection ω to be the holonomy group of the induced principal bundle connection $\overline{\omega}$. Proposition 2.1.1 shows how these different definitions of the holonomy group of a Cartan connection are related.

Therefore,

$$\omega(\dot{\tilde{\gamma}}(t)) = -\operatorname{Ad}\big(g(t)\big)\,\omega_G\big(\dot{g}(t)\big) = -dR_{g^{-1}(t)}\dot{g}(t) = dL_{a(t)^{-1}}\dot{a}(t)\ ,$$

where $a(t) := g(t)^{-1}$. Thus, $a := g^{-1} : [\alpha, \beta] \to G$ is the development of ω along $\tilde{\gamma}$. Using the definition of the holonomy of γ with respect to ω (cf. (2.1.7), (2.1.8) and (2.1.9)) we obtain that $\mathrm{hol}_u^{\omega}(\gamma) = g(\beta)^{-1}$. Thus, $\mathrm{hol}_u^{\omega}(\gamma) = \mathrm{hol}_{\overline{u}}^{\overline{\omega}}(\gamma)$ for all loops $\gamma \in \mathcal{L}(M, x)$ which admit a closed lift $\tilde{\gamma} \in \mathcal{L}(P, u)$.

It remains to show that $\mathcal{L}(M, x) = \pi(\mathcal{L}(P, u))$ if B is connected or if M is simply connected. From the homotopy lifting lemma for fibrations follows that

$$\pi(\mathcal{L}(P, u)) = \{\gamma \in \mathcal{L}(M, x) \mid [\gamma] \in \pi_\sharp(\pi_1(P, u))\}.$$

The exact homotopy sequence of the fibration $(P, \pi, M; B)$ shows that the induced map $\pi_\sharp$ on the fundamental groups is surjective if the structure group B is connected. Therefore, $\mathcal{L}(M, x) = \pi(\mathcal{L}(P, u))$ if the B is connected or if M is simply connected. The statements (2.1.12) and (2.1.13) follow then from the definition of the holonomy groups in question. Statement (2.1.14) is a consequence of (2.1.5). □

The correspondence of the holonomy groups allows us to translate the well-known properties of holonomy groups of connection forms on principal fibre bundles (resp. covariant derivatives) into properties of holonomy groups of Cartan connections. In particular, the holonomy algebra is generated by the curvature.

In general, the *curvature of a 1-form* $\sigma \in \Omega^1(N, \mathfrak{g})$ on a manifold N with values in a Lie algebra $\mathfrak{g}$ is the 2-form $\Omega^\sigma \in \Omega^2(N, \mathfrak{g})$ defined by

$$\Omega^\sigma := d\sigma + \frac{1}{2}[\sigma, \sigma].$$

Hence, the *curvature* Ω^ω *of a Cartan connection* $\omega \in \Omega^1(P, \mathfrak{g})$ is a 2-form on P with values in the Lie algebra $\mathfrak{g}$. The invariance properties of a Cartan connection yield that its curvature is horizontal and of type Ad, meaning that

$$\begin{aligned} TvP \lrcorner\, \Omega^\omega &= 0 \qquad \text{and} \\ R_b^*\Omega^\omega &= \operatorname{Ad}(b^{-1}) \circ \Omega^\omega \qquad \forall\, b \in B, \end{aligned}$$

where $TvP \subset TP$ is the vertical tangent bundle of P. We call ω *flat*, if its curvature is zero. For example, the Maurer-Cartan form ω_G of a Lie group G considered as Cartan connection on the homogeneous bundle $(G, \pi, G/B; B)$ is flat. For this reason, the homogeneous bundle $(G, \pi, G/B; B)$ with the Maurer-Cartan form ω_G is called *the flat model of Cartan geometry of type* G/B. The Uniformization Theorem in Cartan geometry says that – up to discrete groups – this is the only situation where complete flat connections appear. More precisely:

Proposition 2.1.2. (cf. [Sh97], chap. 5.5) *Let $(P, \pi, M; B)$ be a principal fibre bundle with connected structure group B over a connected manifold M which admits a complete flat Cartan connection $\omega \in \Omega^1(P, \mathfrak{g})$. Then, there exists a connected Lie group G with Lie algebra $\mathfrak{g}$, containing B as closed subgroup, such that M is diffeomorphic to the locally homogeneous space $\Gamma \setminus G/B$, where $\Gamma \subset G$ is a discrete subgroup. Moreover, the B-bundle $(\Gamma \setminus G, \mathrm{pr}, \Gamma \setminus \mathrm{G/B}; \mathrm{B})$ with the Cartan connection $\omega_{\Gamma \setminus G}$, given by the Maurer-Cartan form of G, is isomorphic to (P, ω).*

The next proposition shows how the Lie algebra of the holonomy group of a Cartan connection is determined by its curvature.

Proposition 2.1.3. *Let $\omega \in \Omega^1(P, \mathfrak{g})$ be a Cartan connection on the B-principal fibre bundle P over M, and let $u \in P_x$. Then the Lie algebra of the holonomy group $\mathrm{Hol}_u(P, \omega)$ is given by*

$$\mathfrak{hol}_u(P,\omega) = \mathrm{span}\left\{ \mathrm{Ad}\left(\mathrm{hol}^\omega(\lambda)\right) \circ \Omega^\omega_p(X,Y) \;\middle|\; \begin{array}{l} p \in P,\ X, Y \in T_pP\,, \\ \lambda \text{ path in } P \text{ from } u \text{ to } p \end{array} \right\} \subset \mathfrak{g}.$$

Furthermore, if $\rho : G \to GL(V)$ is a G-representation, $E := P \times_B V$ the associated tractor bundle, ∇^ω the covariant derivative on E induced by the Cartan connection ω and $R^{\nabla^\omega}(X,Y) := [\nabla^\omega_X, \nabla^\omega_Y] - \nabla^\omega_{[X,Y]}$ its curvature endomorphism, then

$$\mathfrak{hol}_x(E, \nabla^\omega) = \mathrm{span}\left\{ \mathcal{P}^{\nabla^\omega}_{\gamma^{-1}} \circ R^{\nabla^\omega}_y(X,Y) \circ \mathcal{P}^{\nabla^\omega}_{\gamma} \;\middle|\; \begin{array}{l} X, Y \in T_yM \\ \gamma \text{ path in } M \text{ from } x \text{ to } y \end{array} \right\}.$$

Proof. Let $\overline{\omega}$ be the connection form on the extended G-bundle $\overline{P}$, induced by the Cartan connection ω (cf. formula (2.1.10)). Furthermore, let $\overline{u} := [u, e] \in \overline{P}$. According to (2.1.12), the Lie algebras $\mathfrak{hol}_u(P, \omega)$ and $\mathfrak{hol}_{\overline{u}}(\overline{P}, \overline{\omega})$ coincide. By the Ambrose-Singer Theorem (cf. [KN63], chap. II, Th. 8.1) the holonomy algebra of the connection form $\overline{\omega}$ is given by

$$\mathrm{hol}_{\overline{u}}(\overline{P}, \overline{\omega}) = \mathrm{span}\left\{ \Omega^{\overline{\omega}}_{[p,g]}(\overline{X}, \overline{Y}) \;\middle|\; \begin{array}{l} \exists\ \overline{\omega}\text{-horizontal path in } \overline{P} \text{ from } \overline{u} \text{ to } [p,g]\,, \\ \overline{X}, \overline{Y} \in T_{[p,g]}\overline{P} \end{array} \right\}.$$

Let $\delta(t) = [\lambda(t), g(t)]$ be a path in $\overline{P}$ starting in $\overline{u} = [u, e]$. As in the proof of Proposition 2.1.1 it follows that $\delta(t) = [\lambda(t), g(t)]$ is $\overline{\omega}$-horizontal exactly if λ is an arbitrary path in P starting from u, and that $a(t) := g(t)^{-1}$ is the development of ω along λ. Thus

$$\mathfrak{hol}_{\overline{u}}(\overline{P}, \overline{\omega}) = \mathrm{span}\left\{ \Omega^{\overline{\omega}}_{[p,a^{-1}]}(\overline{X}, \overline{Y}) \;\middle|\; \begin{array}{l} p \in P,\ \ a = \mathrm{hol}^\omega(\lambda) \text{ for a path } \lambda \text{ in } P \\ \text{from } u \text{ to } p\,,\ \ \overline{X}, \overline{Y} \in T_{[p,a^{-1}]}\overline{P} \end{array} \right\}.$$

Using formula (2.1.10), we obtain for the curvatures of the Cartan connection ω on P and the induced connection form $\overline{\omega}$ in $\overline{P}$,

$$\Omega^{\overline{\omega}}_{[p,a^{-1}]}(\overline{X}, \overline{Y}) = \mathrm{Ad}(a)\, \Omega^\omega_p(X, Y)$$

with appropriate $X, Y \in T_pP$. Hence,

$$\begin{aligned}\mathfrak{hol}_u(P,\omega) &= \mathfrak{hol}_{\overline{u}}(\overline{P},\overline{\omega})\\ &= \operatorname{span}\left\{ \operatorname{Ad}\left(\mathrm{hol}^{\omega}(\lambda)\right)\circ \Omega^{\omega}_p(X,Y) \;\middle|\; \begin{array}{l} p\in P,\ X,Y\in T_pP\,,\\ \lambda \text{ path in } P \text{ from } u \text{ to } p\end{array}\right\}.\end{aligned}$$

The second formula in the proposition results from the relation between the curvature form $\Omega^{\overline{\omega}}$ on the principal bundle $\overline{P}$ and the curvature endomorphism $R^{\nabla^{\omega}}$ on E:

$$\rho_*(\Omega^{\overline{\omega}}_q(\overline{X},\overline{Y})) = [q]^{-1}\circ R^{\nabla^{\omega}}_y(X,Y)\circ [q]\,,$$

where $q \in \overline{P}_y$ and $\pi(\overline{X}) = X$, $\pi(\overline{Y}) = Y \in T_yM$. For the holonomy algebra we have to take into account all points q that arise by parallel displacement from u along curves γ connecting x with y in M, i.e., all points q with $[q] = \mathcal{P}^{\nabla^{\omega}}_{\gamma}\circ[u]$. This gives the second statement. □

In special situations there is a more appropriate formula for the holonomy algebra of a Cartan connection. Let us denote by ∇^{End} the covariant derivative in the endomorphism bundle $\mathrm{End}(E)$ induced by ∇^{ω}:

$$(\nabla^{\mathrm{End}}_X F) := [\nabla^{\omega}_X, F] = \nabla^{\omega}_X\circ F - F\circ\nabla^{\omega}_X \quad \text{for } F\in\Gamma(\mathrm{End}(E)),\ X\in\mathfrak{X}(M).$$

Furthermore, we fix an arbitrary covariant derivative D on TM. For a 2-form σ on M with values in $\mathrm{End}(E)$ we define by induction

$$(\widetilde{\nabla}_{V_1}\sigma)(X,Y) := \nabla^{\mathrm{End}}_{V_1}(\sigma(X,Y)) - \sigma(D_{V_1}X,Y) - \sigma(X,D_{V_1}Y)$$

and

$$\begin{aligned}(\widetilde{\nabla}^k_{V_1\dots V_k}\sigma)(X,Y) := \;&\nabla^{\mathrm{End}}_{V_k}((\widetilde{\nabla}^{k-1}_{V_1\dots V_{k-1}}\sigma)(X,Y))\\ &- (\widetilde{\nabla}^{k-1}_{V_1\dots V_{k-1}}\sigma)(D_{V_k}X,Y) - (\widetilde{\nabla}^{k-1}_{V_1\dots V_{k-1}}\sigma)(X,D_{V_k}Y)\\ &- \sum_{i=1}^{k-1}(\widetilde{\nabla}^{k-1}_{V_1\dots D_{V_k}V_i\dots V_{k-1}}\sigma)(X,Y).\end{aligned}$$

Proposition 2.1.4. *Let ∇^{ω} be the covariant derivative on the tractor bundle E defined by the Cartan connection ω, and let $x\in M$. Consider the following subspace in $\mathrm{GL}(E_x)$:*

$$\mathfrak{hol}'_x(E,\nabla^{\omega}) := \operatorname{span}\left\{ (\widetilde{\nabla}^k_{V_1,\dots,V_k}R^{\nabla^{\omega}})_x(X,Y) \;\middle|\; \begin{array}{l} X,Y,V_1,\dots,V_k\in T_xM\,,\\ 0\le k<\infty\end{array}\right\}.$$

If the dimension of $\mathfrak{hol}'_x(E,\nabla^E)$ is constant in x, then this vector space coincides with the holonomy algebra $\mathfrak{hol}_x(E,\nabla^{\omega})$. In particular, this is the case if all data are analytic.

Proof. This is a translation of the corresponding statements for the infinitesimal holonomy group of affine connections in [KN63], vol.1. □

Finally, let us state the basic relation between the holonomy group of a Cartan connection ω and ∇^ω-parallel sections in the associated tractor bundles.

Proposition 2.1.5. *Let ω be a Cartan connection on the B-principal fibre bundle P over M, $E = P \times_B V$ a tractor bundle associated to P, ∇^ω the induced covariant derivative on E and $u \in P_x$ a fixed point. Let B be connected or M be simply connected. Then there is a 1-1 correspondence between the space of parallel sections*

$$\{\varphi \in \Gamma(E) \mid \nabla^\omega \varphi = 0\}$$

and the space of holonomy invariant vectors

$$\{v \in V \mid \rho\big(\mathrm{Hol}_u(P,\omega)\big)\, v = v\} \overset{[u]}{\simeq} \{e \in E_x \mid \mathrm{Hol}_x(E,\nabla^\omega)\, e = e\}.$$

Proof. For $v \in V$ we define a section $\varphi_v \in \Gamma(E)$ by the parallel transport of the element $e = [u,v] \in E_x$ with respect to ∇^ω:

$$\varphi_v(y) := \mathcal{P}_\gamma^{\nabla^\omega}(e) \in E_y \ ,$$

where γ is a curve in M from x to y. The section φ_v does not depend on the choice of γ, φ_v is parallel and the map $v \mapsto \varphi_v$ is bijective between the space of holonomy invariant vectors and the space of parallel sections. □

This proposition shows, on one hand, how useful the knowledge of the holonomy group is if one wants to determine the spaces of parallel sections. The task to solve the differential equation $\nabla^\omega \varphi = 0$ reduces to a problem of linear algebra. On the other hand, if one has a special kind of geometry (for example conformal geometry) which is characterized by the Cartan connection ω, then the holonomy group – by its fixed vectors on a representation space – defines distinguished geometric invariants (∇^ω-parallel object) of the manifold.

2.2 Holonomy groups of conformal structures

In this section we consider special Cartan connections which are associated to conformal manifolds. A conformal manifold of signature (p,q) is a pair (M,c), where M is a smooth manifold of dimension $n = p+q$ and $c = [g]$ is an equivalence class of metrics of signature[2] (p,q). As we already explained in Chapter 1, two metrics g and $\widetilde{g}$ are called equivalent if $\widetilde{g} = fg$ for a positive smooth function f on M. We always assume $n \geq 3$ for the dimension of M.

In (pseudo-)Riemannian geometry, there is a distinguished covariant derivative on the tangent bundle (and a corresponding distinguished connection form

[2]Here p denotes the number of -1 and q the number of $+1$ in the normal form of the metric.

on the frame bundle) of M associated to a metric g, the Levi-Civita connection, which is used to define isometry invariant geometric objects on (M, g). Contrary to that, there can't be a distinguished covariant derivative on the tangent bundle (or a connection form on the frame bundle) of M depending only on the conformal structure $c = [g]$ itself, since one needs derivatives up to second order to determine conformal diffeomorphisms uniquely.

There are several ways to do differential geometry on conformal manifolds. One is to define conformally invariant objects using a metric $g \in c$ and to prove then that the defined object is independent of the choice of the metric $g \in c$. This often leads to long calculations and the geometric meaning of the object sometimes remains in the dark. Another way is to define the geometric objects from the beginning using only the conformal class c and not a single metric. For this purpose one can use the conformally invariant differential calculus for conformal manifolds which is provided by Cartan geometry. In this approach the bundle of frames on M is replaced by a larger bundle $\mathcal{P}^1$, which is obtained from the conformal frame bundle by a process called *first prolongation*. This first prolongation admits a distinguished Cartan connection, the *normal conformal Cartan connection* ω^{nor}, which in conformal geometry plays the same role as the Levi-Civita connection plays in (pseudo)Riemannian geometry. It is used to define conformally invariant differentiations on all tractor bundles defined on conformal manifolds. In analogy to (pseudo)Riemannian geometry, it is then natural to define *the holonomy group of the conformal manifold* (M, c) to be the holonomy group of this normal conformal Cartan connection[3]:

$$\mathrm{Hol}(M, c) := \mathrm{Hol}(\mathcal{P}^1, \omega^{nor}).$$

The aim of this section is to give an introduction into this conformally invariant approach. In particular, we will define the normal conformal Cartan connection and describe the induced covariant derivative in the associated tractor bundles. For further reading and detailed proofs we recommend the book [CS09].

2.2.1 The first prolongation of the conformal frame bundle

First, we want to describe the groups B and G (compare Section 2.1) which are used in conformal geometry. In Section 1.1 these groups were already considered for Riemannian signature. In this case, G is the group of all conformal diffeomorphisms of the sphere S^n with the conformal structure given by the round metric and $B \subset G$ is the stabilizer of a point. In order to fix the notation and sign conventions of this chapter, we will shortly describe the groups G and B and its Lie algebras again for arbitrary signature. For this purpose, let us consider the isotropic cone

$$C^{p,q} := \{x \in \mathbb{R}^{p+1,q+1} \setminus \{0\} \mid \langle x, x \rangle_{p+1,q+1} = 0\}$$

[3]For the precise definition see Definition 2.2.3

in the pseudo-Euclidean space $(\mathbb{R}^{p+1,q+1}, \langle\cdot,\cdot\rangle_{p+1,q+1})$. The projective light cone $Q^{p,q} := \mathbb{P}C^{p,q}$ can be equipped with a conformal structure c of signature (p,q):

$$c := \left[\mu^* \langle\cdot,\cdot\rangle_{p+1,q+1}\right],$$

where $\mu : Q^{p,q} \longrightarrow C_+^{p,q}$ is an arbitrary section from the projectivization into a fixed component $C_+^{p,q}$ of the cone. The conformal manifold $(Q^{p,q}, c)$ is called a *Möbius sphere of signature* (p,q). For Riemannian conformal structures (signature $(0,n)$), the Möbius sphere is diffeomorphic to the sphere S^n equipped with the conformal class of the round metric. Moreover, as we learned in the first chapter, it is the conformal boundary of the hyperbolic space H^{n+1} as well as the conformal compactification of the Euclidean space $\mathbb{R}^n$. For Lorentzian conformal structures (signature $(1,n-1)$), the Möbius sphere is diffeomorphic to $(S^1 \times S^{n-1})/\mathbb{Z}_2$ with the conformal class of the product metric. In physics literature this space appears under the name 'Einstein universe Ein_n'. It is the conformal boundary of the Anti de Sitter space AdS^{n+1} as well as the conformal compactification of the Minkowski space $\mathbb{R}^{1,n-1}$.

Problem 2.2.1. *Prove that there is an embedding* $\iota : \mathbb{R}^{p,q} \longrightarrow Q^{p,q}$, *such that* $\iota(\mathbb{R}^{p,q}) \subset Q^{p,q}$ *is an open, dense set and* $\langle\cdot,\cdot\rangle_{p,q} \in \iota^* c$. *Describe the boundary* $\partial\,\iota(\mathbb{R}^{p,q}) = Q^{p,q} \setminus \iota(\mathbb{R}^{p,q}) \subset Q^{p,q}$ *in the Riemannian and in the Lorentzian case.*

For convenience, we will work in this chapter with the following basis of $\mathbb{R}^{p+1,q+1}$: We fix an orthonormal basis $(e_0, \ldots, e_p, e_{p+1}, \ldots, e_{n+1})$ in $\mathbb{R}^{p+1,q+1}$, where the first $p+1$ vectors are timelike, and denote by f_0 and f_{n+1} the isotropic vectors

$$f_0 := \frac{1}{\sqrt{2}}(e_{n+1} - e_0) \quad \text{and} \quad f_{n+1} := \frac{1}{\sqrt{2}}(e_{n+1} + e_0).$$

Then $\mathbb{R}^{p+1,q+1} = \mathbb{R}f_0 \oplus \mathbb{R}^{p,q} \oplus \mathbb{R}f_{n+1}$, where $\mathbb{R}^{p,q} = \mathrm{span}(e_1, \ldots, e_n)$. In the following we will write all linear maps on $\mathbb{R}^{p+1,q+1}$ as matrices with respect to the basis $(f_0, e_1, \ldots, e_n, f_{n+1})$. The Gram matrix of $\langle\cdot,\cdot\rangle_{p+1,q+1}$ with respect to this basis has the form $\begin{pmatrix} 0 & 0 & 1 \\ 0 & \mathbb{J}_{p,q} & 0 \\ 1 & 0 & 0 \end{pmatrix}$ where $\mathbb{J}_{p,q} = \begin{pmatrix} -\mathbb{I}_p & 0 \\ 0 & \mathbb{I}_q \end{pmatrix}$ and $\mathbb{I}_k$ is the identity matrix with k rows. For a row vector $z \in (\mathbb{R}^n)^*$ let $z^\sharp := \mathbb{J}_{p,q} z^t$. For a column vector $x \in \mathbb{R}^n$ let $x^\flat := x^t \mathbb{J}_{p,q}$.

Let $\mathrm{O}(p+1, q+1)$ be the pseudo-orthogonal group of signature $(p+1, q+1)$. The group G one works with in conformal geometry is the Möbius group, i.e., the group $G := \mathrm{PO}(p+1, q+1) := \mathrm{O}(p+1, q+1)/\{\pm 1\}$. This group is isomorphic to the group $\mathrm{Conf}(Q^{p,q}, c)$ of all conformal diffeomorphisms of the Möbius sphere (cf. the explanation in Section 1.1 for the Riemannian case or [Sch97], Chapter 2). The Möbius group acts transitively and effectively on $Q^{p,q}$. We denote by $B \subset G$ the stabilizer of the isotropic line $p_\infty := \mathbb{R}f_0 \in Q^{p,q}$. Hence, $Q^{p,q} \simeq G/B$. A direct calculation shows that the group B is isomorphic to the following subgroup of the matrix group $\mathrm{O}(p+1,q+1)$ (with respect to the basis $(f_0, e_1, \dots, e_n, f_{n+1})$:

$$B = \left\{ \begin{pmatrix} a^{-1} & v & -\frac{1}{2}a\langle v,v\rangle_{p,q} \\ 0 & A & -aAv^\sharp \\ 0 & 0 & a \end{pmatrix} \;\middle|\; \begin{array}{l} a \in \mathbb{R}^+ \\ A \in \mathrm{O}(p,q) \\ v \in (\mathbb{R}^n)^* \text{ row vector} \end{array} \right\} \subset \mathrm{O}(p+1,q+1).$$

The subgroup

$$B_0 := \left\{ \begin{pmatrix} a^{-1} & 0 & 0 \\ 0 & A & 0 \\ 0 & 0 & a \end{pmatrix} \;\middle|\; \begin{array}{l} a \in \mathbb{R}^+ \\ A \in \mathrm{O}(p,q) \end{array} \right\} \subset B$$

is isomorphic to the linear conformal group

$$\begin{aligned} \mathrm{CO}(p,q) &:= \{A \in \mathrm{GL}(n) \mid \exists\, \theta > 0 : \langle Ax, Ay\rangle_{p,q} = \theta \langle x,y\rangle_{p,q} \;\forall\, x,y \in \mathbb{R}^n\} \\ &\simeq \mathbb{R}^+ \times \mathrm{O}(p,q). \end{aligned}$$

The subgroup

$$B_1 = \left\{ \begin{pmatrix} 1 & v & -\frac{1}{2}\langle v,v\rangle_{p,q} \\ 0 & \mathbb{I}_n & -v^\sharp \\ 0 & 0 & 1 \end{pmatrix} \;\middle|\; v \in (\mathbb{R}^n)^* \text{ row vector} \right\} \subset B$$

is an abelian subgroup of $\mathrm{O}(p+1,q+1)$, which is normal in B. Then $B/B_1 \simeq B_0 \simeq \mathrm{CO}(p,q)$, where the projection from B to B_0 corresponds geometrically to the mapping $f \in \{\phi \in \mathrm{Conf}(Q^{p,q},c) \mid \phi(p_\infty) = p_\infty\} \mapsto df_{p_\infty} \in \mathrm{CO}(T_{p_\infty}Q^{p,q})$.

Next, let us describe the structure of the Lie algebra $\mathfrak{g}$ of G. $\mathfrak{g} = \mathfrak{o}(p+1,q+1)$ is a $|1|$-graded Lie algebra. It decomposes into a sum of subspaces

$$\mathfrak{g} = \mathfrak{b}_{-1} \oplus \mathfrak{b}_0 \oplus \mathfrak{b}_1 \qquad \text{with} \quad [\mathfrak{b}_i, \mathfrak{b}_j]_{\mathfrak{g}} \subset \mathfrak{b}_{i+j},$$

where $\mathfrak{b}_0$ is the Lie algebra of B_0, $\mathfrak{b}_1$ is the Lie algebra of B_1 and $\mathfrak{b} = \mathfrak{b}_0 \oplus \mathfrak{b}_1$ is the Lie algebra of B. In terms of matrices

$$\mathfrak{g} = \left\{ \mathbb{M}(x,(A,a),z) := \begin{pmatrix} -a & z & 0 \\ x & A & -z^\sharp \\ 0 & -x^\flat & a \end{pmatrix} \;\middle|\; \begin{array}{l} x \in \mathbb{R}^n, z \in (\mathbb{R}^n)^* \\ a \in \mathbb{R}, A \in \mathfrak{o}(p,q) \end{array} \right\},$$

$$\begin{aligned} \mathfrak{b}_{-1} &= \{\, \mathbb{M}(x,(0,0),0) \mid x \in \mathbb{R}^n \,\} \simeq \mathbb{R}^n, \\ \mathfrak{b}_1 &= \{\, \mathbb{M}(0,(0,0),z) \mid z \in (\mathbb{R}^n)^* \,\} \simeq \mathbb{R}^{n*}, \\ \mathfrak{b}_0 &= \{\, \mathbb{M}(0,(A,a),0) \mid A \in \mathfrak{o}(p,q), a \in \mathbb{R} \,\} \simeq \mathfrak{co}(p,q). \end{aligned} \tag{2.2.1}$$

For elements $x \in \mathfrak{b}_{-1}$, $(A,a) \in \mathfrak{b}_0$, $z \in \mathfrak{b}_1$ we have

$$[(A,a),x] = ax + Ax \qquad \in \mathfrak{b}_{-1},$$

$$
\begin{aligned}
[(A,a),z] &= -az - zA & \in \mathfrak{b}_1\,, \\
[x,z] &= (xz - z^{\sharp}x^{\flat}, zx) & \in \mathfrak{b}_0.
\end{aligned}
\tag{2.2.2}
$$

We identify $\mathfrak{b}_1$ and $\mathfrak{b}_{-1}^*$ using the $\mathrm{Ad}(B_0)$-equivariant isomorphism ϕ given by the Killing form $B_{\mathfrak{g}}$ of $\mathfrak{g}$,

$$\phi(z)x := \frac{1}{2n} B_{\mathfrak{g}}(x,z)\,, \quad x \in \mathfrak{b}_{-1},\ z \in \mathfrak{b}_1.$$

This corresponds to the standard dual pairing between $\mathbb{R}^n$ and $(\mathbb{R}^n)^*$. The adjoint representation of B_0 on $\mathfrak{b}_{-1}$ yields an isomorphism between B_0 and the group of linear conformal isomorphisms of $(\mathfrak{b}_{-1} = \mathbb{R}^n, \langle\cdot,\cdot\rangle_{p,q})$.

Now, let (M^n, c) be an arbitrary conformal manifold of signature (p,q) and dimension $n = p+q \geq 3$. We call a frame $(x; s_1, \dots, s_n)$ on M a *conformal frame*, if the vectors $(s_1, \dots, s_n)$ form a pseudo-orthonormal basis in T_xM with respect to a metric g in the conformal class c. We denote by $\pi^0 : \mathcal{P}^0 \to M$ the bundle of all conformal frames of (M,c). This bundle is a reduction of the frame bundle of M to the linear conformal group $B_0 = \mathrm{CO}(p,q) \subset \mathrm{GL}(n)$.

The basic B-principal fibre bundle over M, which is used in conformal geometry, is obtained from the conformal frame bundle $\mathcal{P}^0$ by first prolongation. There are several equivalent models for the first prolongation of $\mathcal{P}^0$. We will describe the first prolongation using horizontal subspaces in the tangent space of $\mathcal{P}^0$.

For this purpose, let $u \in \mathcal{P}^0$ be a conformal frame in $x \in M$. As we already explained, by a *horizontal subspace* $H \subset T_u\mathcal{P}^0$ we mean a subspace that is complementary to the vertical tangent space $Tv_u\mathcal{P}^0 \subset T_u\mathcal{P}^0$. To any horizontal subspace $H \subset T_u\mathcal{P}^0$ we can associate a 2-form $t(H) \in \Lambda^2(\mathfrak{b}_{-1}^*) \otimes \mathfrak{b}_{-1}$, called the *torsion of* H, which we want to define now: First, note that the identification between the matrix representation of B_0 on $\mathbb{R}^n$ and the adjoint representation of B_0 on $\mathfrak{b}_{-1}$ allows us to represent the tangent bundle of M in the form

$$TM = \mathcal{P}^0 \times_{(B_0,\mathrm{Ad})} \mathfrak{b}_{-1}.$$

In this representation the displacement form θ (cf. Example 2.1.2) is a 1-form on $\mathcal{P}^0$ with values in $\mathfrak{b}_{-1}$:

$$\theta_u(X) := [u]^{-1} d\pi^0_u(X) \in \mathfrak{b}_{-1} \qquad \text{for } u \in \mathcal{P}^0,\ X \in T_u\mathcal{P}^0.$$

In particular, $\theta_{|H} : H \to \mathfrak{b}_{-1}$ is an isomorphism for any horizontal space $H \subset T_u\mathcal{P}^0$. Thus, we can define the 2-form $t(H) \in \Lambda^2(\mathfrak{b}_{-1}^*) \otimes \mathfrak{b}_{-1}$,

$$t(H)(v,w) := d\theta_{\,u}\Big((\theta_{|H})^{-1}(v), (\theta_{|H})^{-1}(w)\Big) \in \mathfrak{b}_{-1}\,, \quad v, w, \in \mathfrak{b}_{-1}.$$

A horizontal space H is called *torsion free*, if $t(H) = 0$. With this preparation we can define the *first prolongation* $\mathcal{P}^1$ by

$$\mathcal{P}^1 := \{H \subset T_u\mathcal{P}^0 \mid u \in \mathcal{P}^0 \text{ and } H \text{ horizontal and torsion free}\}.$$

There are natural projections

$$\begin{array}{rccccc} \pi : & \mathcal{P}^1 & \xrightarrow{\pi^1} & \mathcal{P}^0 & \xrightarrow{\pi^0} & M . \\ & H \subset T_u\mathcal{P}^0 & \longmapsto & u \in \mathcal{P}^0_x & \longmapsto & x \end{array}$$

The projection $\pi : \mathcal{P}^1 \to M$ is a principal fibre bundle with structure group B, whereas $\pi^1 : \mathcal{P}^1 \to \mathcal{P}^0$ is a principal fibre bundle with structure group B_1. Let us explain the action of B on $\mathcal{P}^1$. Any element $b \in B = B_0 \ltimes B_1$ can be represented by $b = b_0 \cdot \exp(Z)$, where $b_0 \in B_0$ and $Z \in \mathfrak{b}_1$. An element $b_0 \in B_0$ 'transports' the horizontal space $H \subset T_u\mathcal{P}^0$ into the point ub_0 using the right action of B_0 on $\mathcal{P}^0$:

$$H \cdot b_0 := dR_{b_0}(H) \subset T_{ub_0}\mathcal{P}^0. \tag{2.2.3}$$

An element $\exp(Z) \in B_1$ acts on $\mathcal{P}^1$ by 'rotating' H inside $T_u\mathcal{P}^0$:

$$H \cdot \exp(Z) := \{X + \widetilde{[Z, \theta(X)]_{\mathfrak{g}}}(u) \mid X \in H\}. \tag{2.2.4}$$

The bundle $(\mathcal{P}^1, \pi, M; B)$ is called the *first prolongation of the conformal frame bundle* $\mathcal{P}^0$. It is the basic bundle that is used in conformal differential geometry. Here the Cartan connections live, in particular the distinguished normal Cartan connection of conformal geometry. If one wants, one can view the bundle $\mathcal{P}^1$ as a substitute for the collection of all the horizontal spaces given by the Levi-Civita connections of the metrics $g \in c$.

Problem 2.2.2. *Check that $(\mathcal{P}^1, \pi, M; B)$ is indeed a B-principal fibre bundle.*

Problem 2.2.3. *What is the first prolongation $\mathcal{P}^1$ of the Möbius sphere $Q^{p,q} = G/B$ with its canonically given conformal structure ?*

2.2.2 The normal conformal Cartan connection – invariant form

In order to distinguish a special Cartan connection, we first describe a certain affine subspace of the space of all Cartan connections of type G/B on $\mathcal{P}^1$, the space of *admissible* Cartan connections. Let $\omega \in \Omega^1(\mathcal{P}^1, \mathfrak{g})$ be a Cartan connection on $\mathcal{P}^1$. With respect to the grading $\mathfrak{g} = \mathfrak{b}_{-1} \oplus \mathfrak{b}_0 \oplus \mathfrak{b}_1$, the 1-form ω splits into the components

$$\omega = \omega_{-1} \oplus \omega_0 \oplus \omega_1.$$

The invariance properties of a Cartan connection show:

1. $\omega(\widetilde{Z}(H)) = Z \in \mathfrak{b}_1$ for all $Z \in \mathfrak{b}_1$. Hence, ω_{-1} and ω_0 vanish on vertical vectors of $\pi^1 : \mathcal{P}^1 \to \mathcal{P}^0$.
2. For $b_1 = \exp(Z) \in B_1$ it holds that

$$\begin{aligned} R^*_{b_1}\omega &= Ad(b_1^{-1}) \circ \omega \\ &= \mathrm{Ad}(\exp(-Z)) \circ \omega \end{aligned}$$

$$
\begin{aligned}
&= e^{-\operatorname{ad}(Z)} \circ \omega \\
&= \Big(1 - \operatorname{ad}(Z) + \frac{1}{2}\operatorname{ad}(Z)^2\Big) \circ \omega \\
&= \underbrace{\omega_{-1}}_{\in \mathfrak{b}_{-1}} \oplus \underbrace{\big(\omega_0 - [Z,\omega_{-1}]\big)}_{\in \mathfrak{b}_0} \oplus \underbrace{\Big(\omega_1 - [Z,\omega_0] + \frac{1}{2}[Z,[Z,\omega_{-1}]]\Big)}_{\in \mathfrak{b}_1}.
\end{aligned}
$$

In particular, ω_{-1} is B_1-invariant and horizontal in $\pi^1 : \mathcal{P}^1 \to \mathcal{P}^0$, therefore, it is a lift of a 1-form on $\mathcal{P}^0$.

Definition 2.2.1. A Cartan connection $\omega \in \Omega^1(\mathcal{P}^1, \mathfrak{g})$ is called admissible, if ω_{-1} and ω_0 satisfy the following conditions:

1. $\omega_{-1} = (\pi^1)^*\theta$, where θ is the displacement form of $\mathcal{P}^0$.
2. Let $\xi \in T_H\mathcal{P}^1$. Then the element $(\omega_0)_H(\xi) \in \mathfrak{b}_0$ is given by the vertical part of the vector $d\pi^1_H(\xi) \in T_u\mathcal{P}^0 = H \oplus Tv_u\mathcal{P}^0$:

$$
d\pi^1_H(\xi) = \text{horizontal part in } H \;+\; \underbrace{\widetilde{(\omega_0)_H(\xi)(u)}}_{\text{vertical part}} .
$$

Clearly, two admissible Cartan connections differ only by the $\mathfrak{b}_1$-component. Let us note that the curvature form Ω^ω of any admissible Cartan connection ω takes its values in the subalgebra $\mathfrak{b} \subset \mathfrak{g}$. To see this, remember that the 2-form Ω^ω is horizontal with respect to the B-bundle $\pi : \mathcal{P}^1 \to M$, and check for the $\mathfrak{b}_{-1}$-part of the curvature form the formula

$$
(\Omega^\omega_{-1})_H(\xi,\eta) = t(H)(v,w), \qquad \text{where } \; \omega_H(\xi) = v,\; \omega_H(\eta) = w \in \mathfrak{b}_{-1}.
$$

The *normal* conformal Cartan connection is an admissible Cartan connection, which is uniquely determined by an additional 'trace' condition on its curvature. To state this condition briefly, we consider the codifferential operator

$$
\partial^*_\omega : \Omega^2(\mathcal{P}^1, \mathfrak{g}) \longrightarrow \Omega^1(\mathcal{P}^1, \mathfrak{g}),
$$

given by

$$
(\partial^*_\omega \eta)_H(X) := \sum_{i=1}^{n} \Big[v_i^*, \eta\big(X, \omega_H^{-1}(v_i)\big)\Big]_{\mathfrak{g}} \qquad H \in \mathcal{P}^1,\; X \in T_H\mathcal{P}^1,
$$

where $(v_1, \dots, v_n)$ is a basis in $\mathfrak{b}_{-1} \simeq \mathbb{R}^n$ and $(v_1^*, \dots, v_n^*)$ is the dual basis in $\mathfrak{b}_1 \simeq (\mathbb{R}^n)^*$.

Proposition 2.2.1. *There is a unique admissible Cartan connection $\omega \in \Omega^1(\mathcal{P}^1, \mathfrak{g})$ such that its curvature form $\Omega^\omega \in \Omega^2(\mathcal{P}^1, \mathfrak{g})$ satisfies $\partial^*_\omega \Omega^\omega = 0$.*

Proof. Admissible Cartan connections ω satisfy $\Omega^\omega_{-1} = 0$. Furthermore, $\mathfrak{b}_1$ is abelian. Hence, for admissible Cartan connections the condition $\partial^*_\omega \Omega^\omega = 0$ is equivalent to

$$\sum_{i=1}^{n} \Big[v_i^*, \Omega_0^\omega(X, \omega_H^{-1}(v_i)) \Big]_{\mathfrak{g}} = 0 \qquad \forall\, H \in \mathcal{P}^1, X \in T_H\mathcal{P}^1, \tag{2.2.5}$$

where Ω_0^ω denotes the $\mathfrak{b}_0$-part of the curvature. Using the Lie algebra structure of $\mathfrak{g}$ and the formulas for ω_{-1} and ω_0 from Definition 2.2.1, one can show that the condition (2.2.5) determines ω_1 uniquely. For details cf. [K72] Th. IV.4.2, [CSS97], or [Fe05], chap.6. □

Definition 2.2.2. The unique admissible Cartan connection $\omega \in \Omega^1(\mathcal{P}^1, \mathfrak{g})$ satisfying the curvature condition $\partial^*_\omega \Omega^\omega = 0$ is called the normal conformal Cartan connection. We will denote it by ω^{nor}.

We defined the normal conformal Cartan connection starting with a conformal manifold (M, c). In Cartan geometry often the opposite viewpoint is used. One works with an *arbitrary* B-principal fibre bundle $(P, \pi, M; B)$ over a manifold M and a Cartan connection $\omega \in \Omega^1(P, \mathfrak{g})$, where the groups $G = O(p+1, q+1)/\{\pm 1\}$ and $B = G_{p_\infty}$ are as above. In this case, $P^0 := P/B_1$ is a B_0-principal fibre bundle over M. The $\mathfrak{b}_{-1}$-part ω_{-1} of the Cartan connection projects down to a 1-form $\theta \in \Omega^1(P^0, \mathfrak{b}_{-1})$ of type Ad with the vertical tangent bundle TvP^0 as kernel. Such a pair (P^0, θ) defines a reduction of the frame bundle of M to the linear conformal group $CO(p, q) = B_0$, which determines a conformal structure c on M. These two approaches are equivalent. If $\mathcal{P}^1$ is the first prolongation of the conformal frame bundle of (M, c) for this conformal structure c and ω satisfies the normality condition $\partial^*_\omega \Omega^\omega = 0$, then there exists a principal fibre bundle isomorphism $\phi : P \to \mathcal{P}^1$ such that $\phi^* \omega^{nor} = \omega$.

Problem 2.2.4. *What is the normal conformal Cartan connection for the Möbius sphere $Q^{p,q} = G/B$?*

A Cartan connection of type G/B depends by definition only on the Lie algebra $\mathfrak{g}$ of G. Hence, there are several possibilities of choosing G with Lie algebra $\mathfrak{g}$. Since in the conformal case the stabilizer group $B \subset O(p+1, q+1)/\{\pm 1\}$ can be realized also as a closed subgroup of the matrix group $O(p+1, q+1)$, we will view the normal conformal Cartan connection ω^{nor} in the following as a Cartan connection of type $O(p+1, q+1)/B$. In particular, we define the holonomy group of the normal conformal Cartan connection ω^{nor} as a subgroup of the matrix group $O(p+1, q+1)$:

$$\mathrm{Hol}_H(\mathcal{P}^1, \omega^{nor}) \subset O(p+1, q+1), \quad H \in \mathcal{P}^1.$$

2.2.3 The normal conformal Cartan connection – metric form

For many applications it is useful to describe the invariantly defined normal conformal Cartan connection ω^{nor} and the induced covariant derivative on tractor bundles in terms of the metrics g in the conformal class c.

Let us recall again the notation for the various curvature tensors, which we used already in Chapter 1. We will denote by R^g the curvature endomorphism given by the Levi Civita connection ∇^g of g,

$$R^g(X,Y) := \nabla^g_X\nabla^g_Y - \nabla^g_Y\nabla^g_X - \nabla^g_{[X,Y]} ,$$

by Ric^g the Ricci tensor and by $\mathrm{scal}^{\mathrm{g}}$ the scalar curvature

$$\begin{aligned}\mathrm{Ric}^g(X,Y) &:= \mathrm{trace}\,(Z \mapsto R^g(Z,X)Y),\\ \mathrm{scal}^{\mathrm{g}} &:= \mathrm{trace}_g\, \mathrm{Ric}^g \,.\end{aligned}$$

P^g denotes the Schouten tensor[4]

$$\mathsf{P}^g := \frac{1}{n-2}\left(\mathrm{Ric}^g - \frac{1}{2(n-1)}\mathrm{scal}^{\mathrm{g}}\cdot \mathrm{g}\right) .$$

We often consider the Schouten tensor as an endomorphism $\mathsf{P}^g : TM \longrightarrow T^*M$, using the identification $\mathsf{P}^g(X)(Y) := \mathsf{P}^g(X,Y)$. The skew-symmetric derivative of the endomorphism P^g gives the Cotton-York tensor, C^g

$$\mathsf{C}^g(X,Y) := \nabla^g_X(\mathsf{P}^g)(Y) - \nabla^g_Y(\mathsf{P}^g)(X).$$

Finally, W^g is the Weyl tensor considered as a 2-form with values in the g-skew-symmetric endomorphism on TM,

$$\mathsf{W}^g(X,Y) := R^g(X,Y) + X^\flat \otimes \mathsf{P}^g(Y)^\sharp + \mathsf{P}^g(X)\otimes Y - \mathsf{P}^g(Y)\otimes X - Y^\flat \otimes \mathsf{P}^g(X)^\sharp ,$$

where $X^\flat$ for a vector field X denotes the dual 1-form $X^\flat(Z) := g(X,Z)$, and $\mu^\sharp$ for a 1-form μ denotes the dual vector field $\mu(Z) = g(\mu^\sharp, Z)$.

Before we describe the normal conformal Cartan connection in its metric form, we show how torsion-free connection forms on the conformal frame bundle $\mathcal{P}^0$ over (M,c) induce admissible Cartan connections on the first prolongation $\mathcal{P}^1$. Let $A \in \Omega^1(\mathcal{P}^0, \mathfrak{b}_0)$ be a connection form on the conformal frame bundle $\mathcal{P}^0$ and $\theta \in \Omega^1(\mathcal{P}^0, \mathfrak{b}_{-1})$ the displacement form as above. The torsion of A is by definition the 2-form $T^A \in \Omega^2(\mathcal{P}^0, \mathfrak{b}_{-1})$ with

$$T^A := d\theta + [A,\theta].$$

If $T^A = 0$, we call A *torsion free*. The connection form A defines a right invariant distribution of horizontal spaces $H_u := \ker A_u \subset T_u\mathcal{P}^0$. One easily checks that

$$t(H_u)(\theta(X),\theta(Y)) = T^A_u(X,Y), \qquad X,Y \in T_u\mathcal{P}^0.$$

[4]Note that several authors use the other sign for the Schouten tensor.

Therefore, any torsion free connection form A on $\mathcal{P}^0$ defines a B_0-equivariant smooth section

$$\begin{array}{rccc} \sigma^A : & \mathcal{P}^0 & \longrightarrow & \mathcal{P}^1 \\ & u & \longmapsto & \ker A_u \end{array}$$

in the B_1-bundle $\pi^1 : \mathcal{P}^1 \to \mathcal{P}^0$ (and vice versa).

Proposition 2.2.2. *For any torsion-free connection form $A \in \Omega^1(\mathcal{P}^0, \mathfrak{b}_0)$ there exists a unique admissible Cartan connection $\omega^A \in \Omega^1(\mathcal{P}^1, \mathfrak{g})$, such that*

$$(\sigma^A)^* \omega^A = \theta + A. \tag{2.2.6}$$

Proof. Let $\sigma^A : \mathcal{P}^0 \to \mathcal{P}^1$ be the B_0-equivariant section defined by A. We consider the map $\tau : \mathcal{P}^1 \to \mathfrak{b}_1$, given by

$$H = \sigma(\pi^1(H)) \cdot \exp(\tau(H)) \qquad \forall\, H \in \mathcal{P}^1,$$

and define $\omega^A \in \Omega^1(\mathcal{P}^1, \mathfrak{g})$ by

$$\begin{aligned} \omega^A_{\sigma(u)} &:= ((\pi^1)^*\theta)_{\sigma(u)} + ((\pi^1)^* A)_{\sigma(u)} + d\tau_{\sigma(u)} && u \in \mathcal{P}^0, \\ \omega^A_H &:= \mathrm{Ad}(b_1^{-1}) \circ \omega^A_{\sigma(u)} \circ dR_{b_1^{-1}} && H = \sigma(u) \cdot b_1 \in \mathcal{P}^1. \end{aligned}$$

A direct calculation shows that ω^A is an admissible Cartan connection. Obviously, $(\sigma^A)^*\omega^A = \theta + A$. □

Now, let us consider a metric g in the conformal class c. We denote by $\mathcal{P}^g \subset \mathcal{P}^0$ the subbundle of all g-orthonormal frames and by A^g the Levi-Civita connection of g on $\mathcal{P}^g$ (which extends to a torsion-free connection on $\mathcal{P}^0$). The choice of $g \in c$ gives us two data:

1. A reduction of the B-bundle $\mathcal{P}^1$ to the bundle of g-orthonormal frames $\mathcal{P}^g$,

$$\sigma^g : \mathcal{P}^g \overset{i}{\hookrightarrow} \mathcal{P}^0 \overset{\sigma^{A^g}}{\longrightarrow} \mathcal{P}^1.$$

2. An admissible Cartan connection ω^{A^g} on $\mathcal{P}^1$ with $(\sigma^g)^*\omega^{A^g} = \theta + A^g$.

Next, we compare the admissible Cartan connections ω^{nor} and ω^{A^g}. As we saw, both Cartan connections differ only by the $\mathfrak{b}_1$-part. To describe this difference, we fix an orthonormal basis $(e_1, \ldots, e_n)$ in $(\mathfrak{b}_{-1} = \mathbb{R}^n, \langle\cdot,\cdot\rangle_{p,q})$ and denote by $(e_1^*, \ldots, e_n^*)$ the dual basis in $\mathfrak{b}_1 \simeq \mathfrak{b}_{-1}^*$. Let $u = (s_1, \ldots, s_n)$ be a conformal frame in a point $x \in M$. Recall that we identified the tangent bundle TM with the associated bundle $\mathcal{P}^0 \times_{B_0} \mathfrak{b}_{-1}$ in such a way that $s_j = [u, e_j]$. If $u \in \mathcal{P}^g \subset \mathcal{P}^0$, then $(s_1, \ldots, s_n)$ is g-orthonormal.

Proposition 2.2.3. *Let g be a metric in the conformal class c. Then the normal conformal Cartan connection is given by*

$$\omega^{nor}_H(\xi) = \omega^{A^g}_H(\xi) - \sum_{j=1}^{n} \mathsf{P}^g_x(d\pi_H(\xi), s_j)\, e_j^* \qquad \textit{for } \xi \in T_H\mathcal{P}^1, \tag{2.2.7}$$

where $x = \pi(H)$ and $(s_1, \dots, s_n) = \pi^1(H)$. In particular, if $\sigma^g : \mathcal{P}^g \longrightarrow \mathcal{P}^1$ is the $O(p,q)$-reduction of $\mathcal{P}^1$ given by the metric g, then

$$((\sigma^g)^* \omega^{nor})_u(X) = \theta_u(X) + A^g_u(X) - \sum_{j=1}^{n} \mathsf{P}^g_x(d\pi^0_u(X), s_j)\, e_j^* \,, \tag{2.2.8}$$

where $X \in T_u\mathcal{P}^g$, $u = (s_1, \dots, s_n) \in \mathcal{P}^g$ and $x = \pi^0(u)$.

Proof. The calculation of the difference $\omega^{nor} - \omega^{A^g}$, which is a 1-form on $\mathcal{P}^1$ with values in $\mathfrak{b}_1$, is straightforward. Details can be found in [CSS97] or [Fe05], Satz 6.8. Formula (2.2.8) follows from (2.2.6) in Proposition 2.2.2. □

2.2.4 The tractor connection and its curvature

As we explained in Section 2.1, any Cartan connection on $\mathcal{P}^1$ defines a covariant derivative on an associated tractor bundle. In this section we want to express the covariant derivative, induced by the normal conformal Cartan connection, and its curvature in terms of a metric g in the conformal class c.

Let $\rho : \mathrm{O}(p+1, q+1) \to \mathrm{GL}(V)$ be a representation on a vector space V, $E = \mathcal{P}^1 \times_B V$ the associated tractor bundle over the conformal manifold (M, c) and $\nabla^{nor} : \Gamma(E) \to \Gamma(T^*M \otimes E)$ the covariant derivative on E induced by the normal conformal Cartan connection ω^{nor} according to (2.1.11). We fix again a metric g in the conformal class c. As we saw in Section 2.2.3, the choice of g allows us to reduce the bundles $\mathcal{P}^0$ and $\mathcal{P}^1$ to the bundle of g-orthonormal frames $\mathcal{P}^g$. Hence, the vector bundles E, TM, T^*M and the bundle $\mathfrak{so}(TM, g)$ of skew-symmetric endomorphisms on (TM, g) can be expressed as bundles associated to the $\mathrm{O}(p,q)$-bundle $\mathcal{P}^g$:

$$\begin{aligned} E &= \mathcal{P}^g \times_{(\mathrm{O}(p,q),\rho)} V, \\ TM &= \mathcal{P}^g \times_{(\mathrm{O}(p,q),\mathrm{Ad})} \mathfrak{b}_{-1}, \\ T^*M &= \mathcal{P}^g \times_{(\mathrm{O}(p,q),\mathrm{Ad})} \mathfrak{b}_1, \\ \mathfrak{so}(TM, g) &= \mathcal{P}^g \times_{(\mathrm{O}(p,q),\mathrm{Ad})} \mathfrak{so}(p,q). \end{aligned}$$

We denote the covariant derivative in E, induced by the Levi-Civita connection of g, with the (same) symbol $\nabla^g : \Gamma(E) \to \Gamma(T^*M \otimes E)$ (cf. (2.1.4)). Moreover, the representation ρ induces g-dependent maps

$$\begin{aligned} \rho^g : TM &\longrightarrow \mathrm{End}(E, E), \\ \rho^g : T^*M &\longrightarrow \mathrm{End}(E, E), \\ \rho^g : \mathfrak{so}(TM, g) &\longrightarrow \mathrm{End}(E, E). \end{aligned}$$

To define these maps, choose a frame $u \in \mathcal{P}^g_x$ and set

$$\rho^g(\Upsilon)\varphi := \Big[u, \rho_*\Big([u]^{-1}\Upsilon\Big)\, [u]^{-1}\varphi\Big] \in E_x, \tag{2.2.9}$$

where $\varphi \in E_x$ and Υ is a tangent vector, a covector or a skew-symmetric map in x, respectively. It is easy to check that the right-hand side does not depend on the choice of $u \in \mathcal{P}^g_x$. We obtain the following 'metric' formula for the normal tractor derivative $\nabla^{nor} : \Gamma(E) \to \Gamma(T^*M \otimes E)$ and its curvature endomorphism

$$R^{nor}(X,Y) := [\nabla^{nor}_X, \nabla^{nor}_Y] - \nabla^{nor}_{[X,Y]} :$$

Proposition 2.2.4. *Let $\rho : O(p+1,q+1) \to GL(V)$ be a representation and $E := \mathcal{P}^1 \times_B V$ the associated tractor bundle. Then, for any metric $g \in c$, the covariant derivative induced on E by the normal conformal Cartan connection is given by*

$$\nabla^{nor}_X = \nabla^g_X + \rho^g(X) - \rho^g(\mathsf{P}^g(X)), \tag{2.2.10}$$

$$R^{nor}(X,Y) = \rho^g(\mathsf{W}^g(X,Y)) - \rho^g(\mathsf{C}^g(X,Y)). \tag{2.2.11}$$

Proof. Let $\phi \in \Gamma(E)$ be locally represented by $\phi_{|U} = [u,v] = [\sigma^g \circ u, v]$, where $u : U \to \mathcal{P}^g$ is a local smooth section in $\mathcal{P}^g$ and $v : U \to V$ is a smooth function. Then, from the formulas (2.1.4), (2.1.11), (2.2.8) and (2.2.9) we obtain

$$\begin{aligned}
\nabla^{nor}_X \phi &\overset{(2.1.11)}{=} \Big[u, X(v) + \rho_*\Big(\omega^{nor}(d(\sigma^g \circ u)(X))\Big)v\Big] \\
&\overset{(2.2.8)}{=} \Big[u, X(v) + \rho_*\Big(\theta(du(X)) + A^g(du(X)) - \sum_{j=1}^n \mathsf{P}^g(X,s_j)e_j^*\Big)\, v\Big] \\
&\overset{(2.1.4)}{=} \nabla^g_X\phi + \Big[u, \rho_*\Big([u]^{-1}X\Big)v\Big] - \Big[u, \rho_*\Big([u]^{-1}(\mathsf{P}^g(X)\Big)v\Big] \\
&\overset{(2.2.9)}{=} \nabla^g_X\phi + \rho^g(X)\phi - \rho^g(\mathsf{P}^g(X))\phi.
\end{aligned}$$

Inserting (2.2.10) into the formula for the curvature endomorphism R^{nor} gives

$$\begin{aligned}
R^{nor}(X,Y) = R^{\nabla^g}(X,Y) &+ [\rho^g(X), \rho^g(Y)] + [\rho^g(\mathsf{P}^g(X)), \rho^g(\mathsf{P}^g(Y))] \\
&- [\rho^g(X), \rho^g(\mathsf{P}^g(Y))] - [\rho^g(\mathsf{P}^g(X)), \rho^g(Y)] \\
&+ [\nabla^g_X, \rho^g(Y)] + [\rho^g(X), \nabla^g_Y] - \rho^g([X,Y]) \\
&- [\nabla^g_X, \rho^g(\mathsf{P}^g(Y))] - [\rho^g(\mathsf{P}^g(X)), \nabla^g_Y] + \rho^g(\mathsf{P}^g([X,Y])).
\end{aligned}$$

Using (2.2.9) for the endomorphism ρ^g, formula (2.2.2) for the Lie bracket $[\mathfrak{b}_{-1}, \mathfrak{b}_1]_{\mathfrak{g}}$ and taking into account that the subalgebras $\mathfrak{b}_{-1}$ and $\mathfrak{b}_1$ of $\mathfrak{g}$ are abelian, we obtain

$$\begin{aligned}
R^{nor}(X,Y) &= R^{\nabla^g}(X,Y) - [\rho^g(X), \rho^g(\mathsf{P}^g(Y))] + [\rho^g(Y), \rho^g(\mathsf{P}^g(X))] \\
&\quad - \rho^g(\mathsf{C}^g(X,Y)) \\
&= \rho^g(\mathsf{W}^g(X,Y)) - \rho^g(\mathsf{C}^g(X,Y)) .
\end{aligned}$$

□

Now, we apply these formulas to the standard representation ρ, given by the matrix action of $O(p+1, q+1)$ on $\mathbb{R}^{p+1,q+1}$. We denote the associated tractor bundle by

$$\mathcal{T}(M) = \mathcal{P}^1 \times_B \mathbb{R}^{p+1,q+1}$$

and call it the *standard tractor bundle*. The scalar product $\langle \cdot, \cdot \rangle_{p+1,q+1}$ on $\mathbb{R}^{p+1,q+1}$ induces a bundle metric $\langle \cdot, \cdot \rangle$ on $\mathcal{T}(M)$. The normal covariant derivative ∇^{nor} on $\mathcal{T}(M)$ is metric, i.e.,

$$X(\langle \phi, \psi \rangle) = \langle \nabla^{nor}_X \phi, \psi \rangle + \langle \phi, \nabla^{nor}_X \psi \rangle.$$

We fix again a metric g in the conformal class c and reduce $\mathcal{P}^1$ to the frame bundle $\mathcal{P}^g$. If we restrict the representation ρ to the subgroup $O(p, q)$, we obtain the splitting

$$\begin{array}{ccc} \mathbb{R}^{p+1,q+1} & \simeq & \mathbb{R} \oplus \mathbb{R}^{p,q} \oplus \mathbb{R} \\ \alpha f_0 + y + \beta f_{n+1} & \mapsto & (\alpha, y, \beta) \end{array}$$

into three $O(p, q)$-representations, where $O(p, q)$ acts trivially on both 1-dimensional summands and by matrix action on $\mathbb{R}^{p,q}$. Hence, any metric g in the conformal class c defines the following splitting of the standard tractor bundle,

$$\mathcal{T}(M) \overset{g}{\simeq} \underline{\mathbb{R}} \oplus TM \oplus \underline{\mathbb{R}} \tag{2.2.12}$$

where $\underline{\mathbb{R}}$ denotes the trivial line bundle over M. In this identification the bundle metric is given by

$$\langle\, (\alpha_1, Y_1, \beta_1)\, ,\, (\alpha_2, Y_2, \beta_2)\, \rangle = \alpha_1\beta_2 + \alpha_2\beta_1 + g(Y_1, Y_2).$$

Proposition 2.2.5. *If g and $\widetilde{g} = e^{2\varphi} g$ are two metrics in the conformal class c, then the metric representations of a standard tractor transform in the following way:*

$$\begin{array}{ccc} \underline{\mathbb{R}} \oplus TM \oplus \underline{\mathbb{R}} & \overset{g}{\simeq} \mathcal{T}(M) \overset{\widetilde{g}}{\simeq} & \underline{\mathbb{R}} \oplus TM \oplus \underline{\mathbb{R}} \\ \begin{pmatrix} \alpha \\ Y \\ \beta \end{pmatrix} & \longmapsto & \begin{pmatrix} \widetilde{\alpha} \\ \widetilde{Y} \\ \widetilde{\beta} \end{pmatrix} := \begin{pmatrix} e^{-\varphi}\big(\alpha - Y(\varphi) - \frac{1}{2}\beta \|\operatorname{grad}^g \varphi\|_g^2\big) \\ e^{-\varphi}(Y + \beta \operatorname{grad}^g \varphi) \\ e^{\varphi}\beta \end{pmatrix}. \end{array}$$

Proof. Let $u = (s_1, \ldots, s_n)$ and $\widetilde{u} = (e^{-\varphi} s_1, \ldots, e^{-\varphi} s_n)$ be a local orthonormal basis for g and $\widetilde{g}$ over $U \subset M$, respectively. Then $\widetilde{u} = u \cdot b_0$, where

$$b_0 = (e^{-\varphi}, \mathbb{I}_n) = \begin{pmatrix} e^{\varphi} & 0 & 0 \\ 0 & \mathbb{I}_n & 0 \\ 0 & 0 & e^{-\varphi} \end{pmatrix} \in B_0.$$

For the sections σ^g and $\sigma^{\widetilde{g}}$, which define the reduction of $\mathcal{P}^1$ to $\mathcal{P}^g$ and $\mathcal{P}^{\widetilde{g}}$, respectively, one obtains

$$\sigma^{\widetilde{g}}(\widetilde{u}) = \sigma^{\widetilde{g}}(u) \cdot b_0 = \sigma^g(u) \cdot \exp(Z) \cdot b_0,$$

where Z takes its values in $\mathfrak{b}_1$. With the transformation formula for the Levi-Civita connection of $\tilde{g}$ and g and the formulas (2.2.3) and (2.2.4) for the action of B on $\mathcal{P}^1$, one proves $Z = [u]^{-1}d\varphi$. Hence,

$$\sigma^g(u) = \sigma^{\tilde{g}}(\tilde{u}) \cdot \underbrace{\begin{pmatrix} e^{-\varphi} & -e^{\varphi}[u]^{-1}d\varphi & -\frac{1}{2}e^{\varphi}\|d\varphi\|_g^2 \\ 0 & \mathbb{I}_n & [u]^{-1}\operatorname{grad}^g\varphi \\ 0 & 0 & e^{\varphi} \end{pmatrix}}_{=:\,b\,\in\, B}.$$

By definition, the metric representation of a standard tractor ϕ with respect to g is given by

$$\phi|_U = [\sigma^g(u), \alpha f_0 + \sum_{i=1}^n y_i e_i + \beta f_{n+1}] \overset{g}{\longmapsto} \begin{pmatrix} \alpha \\ \sum_{i=1}^n y_i s_i \\ \beta \end{pmatrix}.$$

Now, the proposition follows from the representation of ϕ in both reductions:

$$\begin{aligned} \phi|_U &= [\sigma^g(u), (\alpha f_0 + y + \beta f_{n+1})] \\ &= [\sigma^{\tilde{g}}(\tilde{u}) \cdot b, (\alpha f_0 + y + \beta f_{n+1})] \\ &= [\sigma^{\tilde{g}}(\tilde{u}), \rho(b)(\alpha f_0 + y + \beta f_{n+1})] \\ &= [\sigma^{\tilde{g}}(\tilde{u}), (\tilde{\alpha} f_0 + \tilde{y} + \tilde{\beta} f_{n+1})]. \end{aligned}$$

□

Next, we calculate the metric form of ∇^{nor} on $\mathcal{T}(M)$. Let X be a vector field, μ a 1-form on M and τ a g-skew-symmetric endomorphism on TM. Using (2.2.1) and (2.2.9), we obtain for the endomorphisms $\rho^g(X)$, $\rho^g(\mu)$ and $\rho^g(\tau)$ the following formulas on sections of $\mathcal{T}(M)$, represented as a triple (α, Y, β) of two functions α, β and a vector field Y on M:

$$\rho^g(X)\begin{pmatrix} \alpha \\ Y \\ \beta \end{pmatrix} = \begin{pmatrix} 0 \\ \alpha X \\ -g(X,Y) \end{pmatrix}, \qquad \rho^g(\mu)\begin{pmatrix} \alpha \\ Y \\ \beta \end{pmatrix} = \begin{pmatrix} \mu(Y) \\ -\beta\mu^\sharp \\ 0 \end{pmatrix},$$

$$\rho^g(\tau)\begin{pmatrix} \alpha \\ Y \\ \beta \end{pmatrix} = \begin{pmatrix} 0 \\ \tau(Y) \\ 0 \end{pmatrix}.$$

Using these formulas in Proposition 2.2.4, we obtain for the normal derivative ∇^{nor} on the standard tractor bundle the following

Proposition 2.2.6. *Let g be a metric in the conformal class c and consider the corresponding splitting $\mathcal{T}(M) \overset{g}{\simeq} \underline{\mathbb{R}} \oplus TM \oplus \underline{\mathbb{R}}$ of the standard tractor bundle. The smooth sections of $\mathcal{T}(M)$ are identified with triples (α, Y, β), where α and β are* *smooth functions and* Y *is a smooth vector field on* M. *Then the tractor derivative* ∇^{nor} *induced on* $\mathcal{T}(M)$ *by the normal Cartan connection* ω^{nor} *is given by*

$$\nabla^{nor}_X \begin{pmatrix} \alpha \\ Y \\ \beta \end{pmatrix} = \begin{pmatrix} X(\alpha) - \mathsf{P}^g(X,Y) \\ \nabla^g_X Y + \alpha X + \beta\, \mathsf{P}^g(X)^\sharp \\ X(\beta) - g(X,Y) \end{pmatrix}. \tag{2.2.13}$$

The curvature endomorphism of ∇^{nor} *satisfies*

$$R^{nor}(X_1,X_2) \begin{pmatrix} \alpha \\ Y \\ \beta \end{pmatrix} = \begin{pmatrix} -\mathsf{C}^g(X_1,X_2)Y \\ \mathsf{W}^g(X_1,X_2)Y + \beta\, \mathsf{C}^g(X_1,X_2)^\sharp \\ 0 \end{pmatrix}. \tag{2.2.14}$$

Finally, we give the precise definition of what we mean by the holonomy group of a conformal manifold.

Definition 2.2.3. Let (M,c) be a conformal manifold of dimension ≥ 3. The conformal holonomy group of (M,c) with respect to the point $x \in M$ is the group

$$\mathrm{Hol}_x(M,c) := \mathrm{Hol}_x(\mathcal{T}(M), \nabla^{nor}) \subset \mathrm{O}(\mathcal{T}(M)_x, \langle\cdot,\cdot\rangle_x).$$

2.3 Conformal holonomy and Einstein metrics

The conformal holonomy group is an invariant of the conformal structure. Hence, it is natural to ask what information about the conformal structure is encoded in this holonomy group. In the present section, we shortly review some recent results concerning the relation between the conformal holonomy group of (M,c) and the existence of Einstein metrics in the conformal class c. We only sketch the idea of some of the proofs. For more detailed explanations and proofs we recommend the papers of S. Armstrong [Ar07], [Ar06] (Riemannian case), F. Leitner [L05a], [L07] (arbitrary signature) and T. Leistner [Lei06] (Lorentzian case).

We begin with the following remarkable relation between Einstein metrics and parallel sections in the standard tractor bundle $\mathcal{T}(M)$, which is a direct application of the 'metric' formula for the tractor connection ∇^{nor} on $\mathcal{T}(M)$ (cf. Proposition 2.2.6).

Proposition 2.3.1. *Let* (M,c) *be a conformal manifold of dimension* ≥ 3. *If the conformal class* c *contains an Einstein metric* g, *then there exists a non-trivial section* $\phi \in \mathcal{T}(M)$ *with* $\nabla^{nor}\phi = 0$. *On the other hand, if we suppose that there exists a nontrivial* ∇^{nor}*-parallel section* $\phi \in \mathcal{T}(M)$, *then there is an open dense subset* $\widetilde{M} \subset M$ *and an Einstein metric* g *in the conformal class* $c_{|\widetilde{M}}$. *In both cases, for the scalar curvature* scal^g *of* g *it holds that:*

- $\mathrm{scal}^g > 0 \;\Leftrightarrow\; \phi$ *is timelike.*
- $\mathrm{scal}^g < 0 \;\Leftrightarrow\; \phi$ *is spacelike.*

- $\mathrm{scal}^{\mathrm{g}} = 0 \Leftrightarrow \phi$ *is lightlike.*

Proof. Let $g \in c$ be an Einstein metric. Then $\mathrm{Ric}^g = \frac{\mathrm{scal}^{\mathrm{g}}}{n} g$ and therefore the Schouten tensor P^g is a multiple of the metric g,

$$\mathsf{P}^g = \frac{1}{2n(n-1)} \mathrm{scal}^{\mathrm{g}} \cdot \mathrm{g}.$$

Hence, decomposing $\mathcal{T}(M)$ with respect to g as it was described in the latter section, we obtain

$$\nabla^{nor}_X \begin{pmatrix} \alpha \\ Y \\ \beta \end{pmatrix} = \begin{pmatrix} X(\alpha) - \frac{\mathrm{scal}^{\mathrm{g}}}{2n(n-1)} g(X,Y) \\ \nabla^g_X Y + \alpha X + \frac{\beta \mathrm{scal}^{\mathrm{g}}}{2n(n-1)} X \\ X(\beta) - g(X,Y) \end{pmatrix}.$$

Now, consider the non-trivial section $\phi \overset{g}{=} \begin{pmatrix} -\frac{\mathrm{scal}^{\mathrm{g}}}{2n(n-1)} \\ 0 \\ 1 \end{pmatrix} \in \Gamma(\mathcal{T}(M))$. Since the scalar curvature is constant, we obtain

$$\nabla^{nor}_X \phi = \begin{pmatrix} 0 \\ -\frac{\mathrm{scal}^{\mathrm{g}}}{2n(n-1)} X + \frac{\mathrm{scal}^{\mathrm{g}}}{2n(n-1)} X \\ 0 \end{pmatrix} = \begin{pmatrix} 0 \\ 0 \\ 0 \end{pmatrix}.$$

Furthermore,

$$\langle \phi, \phi \rangle = -\frac{\mathrm{scal}^{\mathrm{g}}}{n(n-1)}.$$

This shows the first statement of the proposition. Next, let us suppose, that there is a non-trivial parallel section $\phi \in \Gamma(\mathcal{T}(M))$. We fix an arbitrary metric $g \in c$. Then with respect to this metric g we represent ϕ as $\phi = (\alpha, Y, \beta)^{\perp}$. The condition $\nabla^{nor}\phi = 0$ implies

$$\begin{aligned} Y &= \mathrm{grad}^g \beta, \\ d\alpha &= -\mathsf{P}^g(\mathrm{grad}^g \beta), \\ \alpha g &= \beta \mathsf{P}^g - \mathrm{Hess}^g(\beta). \end{aligned} \tag{2.3.1}$$

Therefore, $\widetilde{M} := \{x \in M \mid \beta(x) \neq 0\}$ is an open dense subset in M, otherwise ϕ would vanish on an open set, hence on M, since ϕ is parallel. Now, consider the metric $\widetilde{g} := \beta^{-2} g$ on $\widetilde{M}$. With the general transformation formulas (cf. [Be87], p.58) and (2.3.1) we obtain for the Schouten tensor of $\widetilde{g}$ on $\widetilde{M}$,

$$\begin{aligned} \mathsf{P}^{\widetilde{g}} &= \mathsf{P}^g - \frac{1}{2}\beta^{-2}|d\beta|^2 g + \beta^{-1} \mathrm{Hess}^g \beta \\ &= -(\alpha\beta + \frac{1}{2}|d\beta|^2)\, \widetilde{g}. \end{aligned}$$

Hence, $\widetilde{g}$ is an Einstein metric and the scalar curvature is given by

$$\frac{1}{2n(n-1)}\mathrm{scal}^{\widetilde{g}} = -\alpha\beta - \frac{1}{2}|d\beta|^2 = -\alpha\beta - \frac{1}{2}g(Y,Y) = -\frac{1}{2}\langle\phi,\phi\rangle.$$

This concludes the proof. □

As application, let us consider the following special case of reduced conformal holonomy.

Proposition 2.3.2. *Let (M, c) be a simply connected conformal manifold of dimension ≥ 3, and let us suppose that there is a 1-dimensional $\mathrm{Hol}_x(M, c)$-invariant subspace $V^1 \subset \mathcal{T}(M)_x$. Then, if V^1 is non-degenerate, there exists an open dense set $\widetilde{M} \subset M$ and an Einstein metric $g \in c|_{\widetilde{M}}$. Thereby, $\mathrm{scal}^g > 0$, if V^1 is timelike, and $\mathrm{scal}^g < 0$, if V^1 is spacelike. If V^1 is lightlike, then there is a Ricci flat metric in c at least locally outside a singular set.*

Proof. We define the line bundle $\mathcal{V} \subset \mathcal{T}(M)$ by parallel displacement of V^1 along M. Since M is simply connected, $\mathcal{V}$ is oriented. (The first Stiefel-Whitney class $w_1(\mathcal{V}) \in H^1(M, \mathbb{Z}_2) = 0$ vanishes). Hence $\mathcal{V}$ admits a global nowhere-vanishing section $\phi \in \Gamma(\mathcal{V})$ with $\mathrm{sgn}(\langle\phi,\phi\rangle) = \mathrm{const}$. Since the bundle $\mathcal{V}$ is parallel, ϕ is recurrent, i.e., there is a 1-form σ on M such that

$$\nabla^{nor}_X \phi = \sigma(X)\,\phi. \tag{2.3.2}$$

If V^1 is space- or timelike, i.e., $\langle\phi,\phi\rangle \neq 0$, the section

$$\psi := \frac{1}{\sqrt{|\langle\phi,\phi\rangle|}} \in \Gamma(\mathcal{T}(M))$$

satisfies $\nabla^{nor}\psi = 0$. Hence we can apply Proposition 2.3.1. If V^1 is lightlike, i.e., $\langle\phi,\phi\rangle = 0$, we proceed as follows. We choose an arbitrary metric $g \in c$ and consider $\phi = \left(\begin{smallmatrix}\alpha\\ Y\\ \beta\end{smallmatrix}\right)$ with respect to the splitting of $\mathcal{T}(M)$ given by g. Then, from (2.3.2) follow

$$\begin{aligned} \beta\,\sigma &= d\beta - Y^b, \\ \sigma(X)\,Y &= \nabla^g_X Y + \alpha\,X + \beta\,\mathsf{P}^g(X,\cdot), \\ \alpha\,\sigma &= d\alpha - \mathsf{P}^g(\cdot,Y). \end{aligned} \tag{2.3.3}$$

These formulas show that $\hat{M} := \{x \in M \mid \beta(x) \neq 0\}$ is open and dense (otherwise, ϕ would vanish on an open set). Now, we restrict our considerations to $\hat{M}$ and suppose w.l.o.g., that $\phi \in \Gamma(\mathcal{T}(\hat{M}))$ is chosen so that $\beta = 1$. Then (2.3.3) yields $d\sigma = 0$. Therefore, for any point $y \in \hat{M}$ we have a neighborhood $U \subset \hat{M}$ and a function $f \in C^\infty(U)$ with $df = \sigma|U$. Now, the local section $\psi := e^{-f}\phi \in \Gamma(\mathcal{T}(M)|_U)$ satisfies $\nabla^{nor}\psi = 0$ and we can apply the previous proposition again. □

The next remarkable fact concerns the relation between conformal and metric holonomy groups of Einstein spaces. It turns out that the conformal holonomy groups of an Einstein space coincides with the metric holonomy group of a certain Ricci-flat ambient metric space. Let us explain this in more detail.

Proposition 2.3.3. *Let (M, g) be an Einstein space of signature (p, q) and dimension $n = p+q \geq 3$ with scalar curvature $\mathrm{scal}^g \neq 0$, and let $(C(M), g_C)$ denote the cone $C(M) := M \times \mathbb{R}^+$ over (M, g) with the Ricci-flat metric*

$$(g_C)_{(x,t)} := \mathrm{sgn}(\mathrm{scal}^g)\Big(\frac{\mathrm{scal}^g}{n(n-1)}\, t^2 g_x + dt^2\Big).$$

Furthermore, let $(\widetilde{M}, \widetilde{g})$ be the ambient metric space

$$\widetilde{M} := \mathbb{R} \times M \times \mathbb{R}^+ \quad \textit{with} \quad \widetilde{g}_{(s,x,t)} := \mathrm{sgn}(\mathrm{scal}^g)\Big(\frac{\mathrm{scal}^g}{n(n-1)}\, t^2 g_x + dt^2 - ds^2\Big).$$

Then, $(\widetilde{M}, \widetilde{g})$ is a Ricci flat pseudo-Riemannian manifold of signature $(p+1, q+1)$, and the holonomy groups satisfy

$$\mathrm{Hol}_x(M, [g]) = \mathrm{Hol}_{(1,x,1)}(\widetilde{M}, \widetilde{g}) = \mathrm{Hol}_{(x,1)}(C(M), g_C).$$

Proof. The Ricci-flatness of $(\widetilde{M}, \widetilde{g})$ follows by calculation from the Einstein condition for g, the signature of $(\widetilde{M}, \widetilde{g})$ is obviously $(p+1, q+1)$. Using the Einstein metric g, we decompose the tractor bundle of $(M, [g])$ in the parts $\mathcal{T}(M) = \underline{\mathbb{R}} \oplus TM \oplus \underline{\mathbb{R}}$ and consider the bundle isomorphism

$$\begin{array}{rccl} F: & \mathcal{T}(M) & \longrightarrow & T\widetilde{M}|_{\{1\}\times M \times \{1\}} \\ & \begin{pmatrix} \alpha \\ Y \\ \beta \end{pmatrix} & \longmapsto & Y + \Big(\frac{\beta\, \mathrm{scal}^g}{2n(n-1)} + \alpha\Big)\frac{\partial}{\partial t} + \Big(\frac{\beta\, \mathrm{scal}^g}{2n(n-1)} - \alpha\Big)\frac{\partial}{\partial s} \end{array}.$$

One can check, that

$$F(\nabla^{nor}_X \phi) = \nabla^{\widetilde{g}}_X(F(\phi)) \quad \text{for all } X \in TM.$$

Therefore, $\mathrm{Hol}_x(M, [g]) = \mathrm{Hol}_{(1,x,1)}(T\widetilde{M}|_{\{1\}\times M\times\{1\}}, \nabla^{\widetilde{g}})$. If $\delta = (\delta_1, \gamma, \delta_2)$ is a curve in $\widetilde{M}$, which is closed in $(1, x, 1)$, then the parallel displacement $\mathcal{P}^{\widetilde{g}}_{\delta} : T\widetilde{M}_{(1,x,1)} \to T\widetilde{M}_{(1,x,1)}$ depends only on the curve $(1, \gamma, 1)$ in $\{1\} \times M \times \{1\}$, i.e., $\mathcal{P}^{\widetilde{g}}_{\delta} = \mathcal{P}^{\widetilde{g}}_{(1,\gamma,1)}$. This shows, that $\mathrm{Hol}_{(1,x,1)}(T\widetilde{M}|_{\{1\}\times M\times\{1\}}, \nabla^{\widetilde{g}}) = \mathrm{Hol}_{(1,x,1)}(\widetilde{M}, \widetilde{g})$. The second statement for the holonomy follows since the vector field $\frac{\partial}{\partial s}$ is parallel on $(\widetilde{M}, \widetilde{g})$. □

For Einstein spaces with scalar curvature $\mathrm{scal}^g = 0$ we can define the ambient metric space in a simpler way:

Proposition 2.3.4. *Let (M,g) be a Ricci-flat Einstein space of signature (p,q) and dimension $n=p+q\geq 3$. We consider the pseudo-Riemannian space $(\overline{M},\overline{g})$ with*

$$\overline{M} := \mathbb{R}\times M\times \mathbb{R}^+ \quad \text{and} \quad \overline{g}_{(s,x,t)} := 2ds\,dt + t^2\, g_x.$$

Then $(\overline{M},\overline{g})$ is a Ricci-flat manifold of signature $(p+1,q+1)$, and the holonomy groups satisfy

$$\mathrm{Hol}_x(M,[g]) = \mathrm{Hol}_{(1,x,1)}(\overline{M},\overline{g}) \subset \mathrm{Hol}_x(M,g)\ltimes \mathbb{R}^n.$$

Proof. The idea of the proof is the same as in the previous proposition. We split $\mathcal{T}(M)$ with respect to the metric g and observe that the bundle map

$$\begin{array}{rccc} F: & \mathcal{T}(M) & \longrightarrow & T\overline{M}|_{\{1\}\times M\times\{1\}} \\ & \begin{pmatrix}\alpha\\ Y\\ \beta\end{pmatrix} & \longmapsto & Y+\alpha\frac{\partial}{\partial t}+\beta\frac{\partial}{\partial s} \end{array}$$

is an isomorphism, which commutes with ∇^{nor} and $\nabla^{\overline{g}}$. Moreover, also in this case, $\mathrm{Hol}_{(1,x,1)}(\overline{M},\overline{g})$ is already generated by parallel translations along curves in $\{1\}\times M\times\{1\}$. Hence, $\mathrm{Hol}_{(1,x,1,)}(M,[g]) = \mathrm{Hol}_{(1,x,1)}(\overline{M},\overline{g})$. For the second statement, we consider a closed curve $\delta=(1,\gamma,1)$ and compare the parallel displacements with respect to $\overline{g}$ and g. In the decomposition $T\overline{M}_{(1,x,1)} = \mathbb{R}\frac{\partial}{\partial s}\oplus T_xM\oplus\mathbb{R}\frac{\partial}{\partial t}$, a direct calculation shows

$$\mathcal{P}^{\overline{g}}_{\delta} = \begin{pmatrix}1 & * & *\\ 0 & \mathcal{P}^g_{\gamma} & *\\ 0 & 0 & 1\end{pmatrix} \subset \mathrm{O}(T\overline{M}_{(1,x,1)},\overline{g}). \tag{2.3.4}$$

□

Problem 2.3.1. *Calculate the parallel transport of the metrics considered in the previous two propositions and verify the statements.*

Another important fact concerns the conformal holonomy group of products of Einstein spaces with appropriate scalar curvature. This group decomposes into the product of the conformal holonomy groups of the factors.

Proposition 2.3.5. *Let (M_1,g_1) and (M_2,g_2) be two Einstein spaces of dimension $m_1\geq 3$ and $m_2\geq 3$, respectively, such that*

$$\lambda := \frac{\mathrm{scal}^{g_1}}{m_1(m_1-1)} = -\frac{\mathrm{scal}^{g_2}}{m_2(m_2-1)} > 0,$$

and let $\big(C(M_1),(g_1)_C\big)$ and $\big(C(M_2),(g_2)_C\big)$ denote the cones defined in Proposition 2.3.3. Then the product of both cones

$$\widetilde{M} := C(M_1)\times C(M_2)$$

with the product metric

$$\widetilde{g} := (g_1)_C \times (g_2)_C = dt_1^2 - dt_2^2 + \lambda t_1^2 g_1 + \lambda t_2^2 g_2$$

is a Ricci-flat space of signature $\mathrm{sgn}(M_1)+\mathrm{sgn}(M_2)+(1,1)$*. The conformal holonomy group of the product of the Einstein spaces* (M_1, g_1) *and* (M_2, g_2) *satisfies*

$$\begin{aligned}
&\mathrm{Hol}_{(x,y)}(M_1 \times M_2, [g_1 \times g_2]) \\
&= \mathrm{Hol}_{(x,1,y,1)}(\widetilde{M}, \widetilde{g}) \\
&= \mathrm{Hol}_{(x,1)}\big(C(M_1), (g_1)_C\big) \times \mathrm{Hol}_{(y,1)}\big(C(M_2), (g_2)_C\big) \\
&= \mathrm{Hol}_x(M_1, [g_1]) \times \mathrm{Hol}_y(M_2, [g_2]).
\end{aligned}$$

Proof. Let $M = M_1 \times M_2$ be the product manifold with the conformal structure $c = [g_1 \times g_2]$. The first equality of the holonomy statements can be shown as in Propositions 2.3.3 and 2.3.4. Using the decomposition of the tractor bundle $\mathcal{T}(M)$ given by the metric $g = g_1 \times g_2 \in c$, one defines a bundle isomorphism from the tractor bundle $\mathcal{T}(\widetilde{M})$ to $T\widetilde{M}|_{M_1\times\{1\}\times M_2\times\{1\}}$, which commutes with ∇^{nor} and $\nabla^{\widetilde{g}}$. Then one shows that $Hol_{(x,1,y,1)}(\widetilde{M}, \widetilde{g})$ is already determined by parallel displacements along curves in $M_1 \times \{1\} \times M_2 \times \{1\}$. The second equality follows from standard holonomy theory of metrics and the third equality comes from Proposition 2.3.3. For details we refer to [L07]. □

The previous propositions show that the conformal holonomy group of an Einstein space can be calculated by the metric holonomy of an ambient pseudo-Riemannian manifold. We note that the classification of holonomy groups of pseudo-Riemannian manifolds is not completed yet. For an overview on known results we refer to [GL08]. Furthermore, let us remark that the ambient metric spaces, which we defined for Einstein spaces or its products, are special cases of the *Fefferman-Graham ambient metric construction* which is explained in Section 1.3. The reader can find more about this in [FG07] and [L07].

Next, we will come to a surprising kind of local de Rham Splitting Theorem for conformal holonomy groups. Let us shortly recall the de Rham Splitting Theorem for (pseudo-)Riemannian manifolds.

Theorem 2.3.1. *Let* (M, g) *be an* n*-dimensional semi-Riemannian manifold, such that there exists a* k*-dimensional non-degenerate* $\mathrm{Hol}_x(M, g)$*-invariant subspace* $E^k \subset T_xM$*. Then,* (M, g) *is locally isometric to a product of semi-Riemannian manifolds, i.e., for any point* $y \in M$ *there exists a neighborhood* $U(y)$ *and two semi-Riemannian manifolds* (U_1, g_1) *and* (U_2, g_2) *of dimension* k *and* $n-k$*, respectively, such that* $(U(y), g)$ *is isometric to* $(U_1, g_1) \times (U_2, g_2)$*. Furthermore,* $\mathrm{Hol}(U(y), g) \simeq \mathrm{Hol}(U_1, g_1) \times \mathrm{Hol}(U_2, g_2)$*. If* (M, g) *is simply connected and geodesically complete, then the splitting is global, i.e.,* (M, g) *is isometric to the product of two semi-Riemannian manifolds* (M_1, g_1) *and* (M_2, g_2) *of dimension* k *and* $n-k$*, respectively.*

A proof of this theorem can be found in [Ba09], Kapitel 5. A similar Splitting Theorem for conformal holonomy groups was proved by St. Armstrong ([Ar07], [Ar06]) in the Riemannian case and by another method by F. Leitner ([L05a], [L07]) for arbitrary signature.

Theorem 2.3.2. *Let* (M, c) *be a simply connected conformal manifold of dimension* $n \geq 3$ *and signature* (p, q) *and suppose that there exists a* k*-dimensional non-degenerate* $\mathrm{Hol}_x(M, c)$*-invariant subspace* $V^k \subset T(M)_x$, *where* $2 \leq k \leq n$. *Then, any point* y *of a certain open dense subset* $\widetilde{M} \subset M$ *has a neighborhood* $U(y)$ *with a metric* $g \in c|_{U(y)}$, *such that* $(U(y), g)$ *is isometric to a product* $(N_1, g_1) \times (N_2, g_2)$, *where* (N_i, g_i) *are Einstein spaces of dimension* $k-1$ *and* $n-(k-1)$, *respectively. If* $k \neq 2, n$, *then the scalar curvatures satisfy*

$$\mathrm{scal}^{g_1} = -\frac{(k-1)(k-2)}{(n-k+1)(n-k)} \mathrm{scal}^{g_2} \neq 0.$$

Moreover, if $\dim N_i \geq 3$, *then*

$$\mathrm{Hol}(U(p), c) \simeq \mathrm{Hol}(N_1, [g_1]) \times \mathrm{Hol}(N_2, [g_2]).$$

2.4 Classification results for Riemannian and Lorentzian conformal holonomy groups

In this section we give a short overview on classification results for conformal holonomy groups in the Riemannian and in the Lorentzian case.

First, we consider simply connected *Riemannian* conformal structures. In the Riemannian case the conformal holonomy group $\mathrm{Hol}(M^n, c)$ is a subgroup[5] of $\mathrm{SO}^0(1, n+1)$. A crucial result is the following algebraic fact:

Proposition 2.4.1. ([DO01]) *Let* $H \subset \mathrm{SO}(1, n+1)$ *be a connected Lie group which acts irreducibly on* $\mathbb{R}^{1,n+1}$. *Then* $H = \mathrm{SO}^0(1, n+1)$.

Hence, if the conformal holonomy group of a simply connected Riemannian conformal manifold (M^n, c) is not the full group $\mathrm{SO}^0(1, n+1)$, then it admits a holonomy invariant subspace. But then, as we saw previously, c contains an Einstein metric or a product of Einstein metrics (at least locally up to a singular set). Therefore, the essential step in the classification of *Riemannian* conformal holonomy groups is the classification of conformal holonomy groups of Einstein spaces which do not decompose conformally. We will call a manifold (M^n, g) *conformally decomposable*, if the conformal class $[g]$ is (possibly outside a singular set) locally represented by a product metric. Otherwise we call (M, g) *conformally indecomposable.* Any 3-dimensional Einstein space (M^3, g) is conformally flat, hence $\mathrm{Hol}^0(M^3, [g]) = 1$.

[5] More precisely, it is a class of conjugated subgroups.

Theorem 2.4.1. ([Ar07], [Ar06]) *Let* (M^n, g) *be a simply connected, conformally indecomposable Riemannian Einstein space of dimension* $n \geq 4$, *and let* $\mathrm{Hol}(M, [g]) \subset \mathrm{SO}^0(1, n+1)$ *be its conformal holonomy group.*

1. *If* $\mathrm{scal}^g < 0$, *then* $\mathrm{Hol}(M, [g]) = \mathrm{SO}^0(1, n)$.
2. *If* $\mathrm{scal}^g > 0$, *then* $\mathrm{Hol}(M, [g])$ *is one of the groups:* $\mathrm{SO}(n+1)$, $\mathrm{SU}(\frac{n+1}{2})$, $\mathrm{Sp}(\frac{n+1}{4})$, G_2 *if* $n = 6$, *or* $\mathrm{Spin}(7)$ *if* $n = 7$.
3. *If* $\mathrm{scal}^g = 0$, *then* $\mathrm{Hol}(M^n, [g])$ *is one of the groups:* $\mathrm{SO}(n) \ltimes \mathbb{R}^n$, $\mathrm{SU}(\frac{n}{2}) \ltimes \mathbb{R}^n$, $\mathrm{Sp}(\frac{n}{4}) \ltimes \mathbb{R}^n$, $\mathrm{G}_2 \ltimes \mathbb{R}^7$ *if* $n = 7$, *or* $\mathrm{Spin}(7) \ltimes \mathbb{R}^8$ *if* $n = 8$.

All groups in the list can be realized as conformal holonomy group of an Einstein space.

Proof. The proof is an application of the previous propositions. First, let (M^n, g) be an Einstein space of $\mathrm{scal}^g > 0$. By Proposition 2.3.1, there exists a 1-dimensional timelike subspace $V^1 \subset \mathbb{R}^{1,n+1}$ on which $\mathrm{Hol}(M, [g])$ acts trivially. The orthogonal complement $(V^1)^\perp \subset \mathbb{R}^{1,n+1}$ is holonomy invariant and spacelike. Moreover, since (M, g) is conformally indecomposable, $\mathrm{Hol}(M, [g])$ acts irreducibly on $(V^1)^\perp$ (Theorem 2.3.2). But, by Proposition 2.3.3, $\mathrm{Hol}(M, [g]) \simeq \mathrm{Hol}(C(M), g_C)$, where $(C(M), g_C)$ is the Ricci-flat Riemannian cone over (M, g). A locally symmetric Riemannian manifold is locally isometric to a symmetric manifold ([H78], chapt. IV.5) and any Ricci-flat symmetric space is flat ([Be87], Thm 7.61). Therefore, we can exclude that the cone is locally symmetric. Hence, $\mathrm{Hol}(M, [g])$ is one of the holonomy groups of Berger's holonomy list of irreducible, non-locally symmetric spaces, which can be realized by Ricci-flat metrics, i.e., $\mathrm{Hol}(M, [g]) = \mathrm{SO}(n+1)$, $\mathrm{SU}(\frac{n+1}{2})$, $\mathrm{Sp}(\frac{n+1}{4})$, G_2 if $n = 6$, or $\mathrm{Spin}(7)$ if $n = 7$. Now, consider the case $\mathrm{scal}^g < 0$. Then there exists a 1-dimensional spacelike subspace $W^1 \subset \mathbb{R}^{1,n+1}$ on which $\mathrm{Hol}(M, [g])$ acts trivially (Proposition 2.3.1). The orthogonal complement $(W^1)^\perp \subset \mathbb{R}^{1,n+1}$ is holonomy invariant and of signature $(1, n)$. There can not be a non-degenerate $\mathrm{Hol}(M, [g])$-invariant subspace in $(W^1)^\perp$, since otherwise (M, g) would be conformally decomposable by Theorem 2.3.2. Moreover, if $W' \subset (W^1)^\perp$ is a degenerate $\mathrm{Hol}(M, [g])$-invariant subspace, $L := W' \cap (W')^\perp$ is a lightlike holonomy-invariant line. Hence, locally, there exists a Ricci-flat Einstein metric conformally equivalent to g (Proposition 2.3.2). But then, (M, g) is conformally decomposable again (cf. [K88]). This shows that $\mathrm{Hol}(M, [g])$ acts irreducibly on $(W^1)^\perp$. Hence, by Proposition 2.4.1, $\mathrm{Hol}(M, [g]) = \mathrm{SO}^0(1, n)$. Finally, let $\mathrm{scal}^g = 0$. By Proposition 2.3.4, $\mathrm{Hol}(M, [g]) \subset \mathrm{Hol}(M, g) \ltimes \mathbb{R}^n$. By the same argument as above, there can not be a non-degenerate holonomy invariant subspace of $\mathbb{R}^{1,n+1}$. Therefore, $\mathrm{Hol}(M, [g])$ acts weakly irreducibly. But then, $\{1\} \ltimes \mathbb{R}^n \subset \mathrm{Hol}(M, [g])$ (cf. [BI93]). Formula (2.3.4) shows that also $\mathrm{Hol}(M, g) \subset \mathrm{Hol}(M, [g])$. Therefore, $\mathrm{Hol}(M, [g]) = \mathrm{Hol}(M, g) \ltimes \mathbb{R}^n$. Since (M, g) is conformally indecomposable, $\mathrm{Hol}(M, g)$ acts irreducibly on $\mathbb{R}^n$. Hence, $\mathrm{Hol}(M, g)$ is one of the groups in Berger's holonomy list of irreducible, Ricci-flat, non-locally symmetric spaces, i.e., $\mathrm{Hol}(M, g) = \mathrm{SO}(n)$, $\mathrm{SU}(\frac{n}{2})$, $\mathrm{Sp}(\frac{n}{4})$, G_2 if $n = 7$, or $\mathrm{Spin}(7)$ if $n = 8$. □

Now, we consider the Lorentzian case. Let $(M^{1,n-1}, c)$ be a simply connected *Lorentzian* conformal manifold. Then the conformal holonomy group $\mathrm{Hol}(M, c)$ is contained in $\mathrm{SO}^0(2, n)$. If $\mathrm{Hol}(M, c)$ acts non-irreducibly, then – possibly outside a singular set – c can be locally represented by a non-Ricci-flat Einstein metric (1-dimensional non-degenerate invariant subspace), by a product of two Einstein metrics with related scalar curvatures (k-dimensional non-degenerate invariant subspace, $2 \leq k \leq n$), by a Ricci-flat Einstein metric (degenerate non-totally isotropic invariant subspace), or by a metric with totally isotropic Ricci tensor and recurrent lightlike vector field (2-dimensional totally isotropic invariant subspace, cf. [Lei06]). For the irreducible case, let us mention the following result:

Proposition 2.4.2. ([DL08]) *Let $H \subset \mathrm{SO}(2, n)$ be a connected Lie group which acts irreducibly on $\mathbb{R}^{2,n}$. Then H is conjugated to one of the following groups:*

1. $\mathrm{SO}^0(2, n)$,
2. $\mathrm{U}(1, \frac{n}{2})$, $\mathrm{SU}(1, \frac{n}{2})$, *if n is even,*
3. $S^1 \cdot \mathrm{SO}^0(1, \frac{n}{2}) \subset \mathrm{U}(1, \frac{n}{2})$, *if n is even and $n \geq 4$,*
4. $\mathrm{SO}^0(1, 2) \subset \mathrm{SO}(2, 3)$, *if $n = 3$.*

It is not known yet whether the group $\mathrm{SO}^0(1, 2) \subset \mathrm{SO}(2, 3)$ can be realized as a holonomy group of a 3-dimensional conformal manifold $(M^{1,2}, c)$. In [L07] and [L08] it is shown that, in case $\mathrm{Hol}^0(M, c) \subset \mathrm{U}(1, \frac{n}{2})$, the group $\mathrm{Hol}^0(M, g)$ is already a subgroup of $\mathrm{SU}(1, \frac{n}{2})$. Since $S^1 \cdot \mathrm{SO}^0(1, \frac{n}{2}) \not\subset \mathrm{SU}(1, \frac{n}{2})$, this group can not be realized as a connected holonomy group of a conformal Lorentzian manifold. We conclude

Theorem 2.4.2. ([DL08]) *Let $(M^{1,n-1}, c)$ be a conformal Lorentzian manifold of dimension $n \geq 4$. If $\mathrm{Hol}^0(M, c)$ acts irreducibly on $\mathbb{R}^{2,n}$, then $\mathrm{Hol}^0(M, g) = \mathrm{SO}^0(2, n)$ or $\mathrm{Hol}^0(M, c) = \mathrm{SU}(1, \frac{n}{2})$ (if n is even).*

$\mathrm{Hol}^0(M, g) = \mathrm{SO}^0(2, n)$ is the generic conformal Lorentzian holonomy group. In Section 2.7 we will describe Lorentzian manifolds with conformal holonomy group $\mathrm{SU}(1, \frac{n}{2})$. Such manifolds can be viewed as a conformal analog of Calabi-Yau manifolds.

2.5 Conformal holonomy and conformal Killing forms

Let (E, ∇^E) be a vector bundle with covariant derivative. Let us suppose that $T^*M \otimes E$ splits into a sum of vector bundles,

$$T^*M \otimes E \simeq E_1 \oplus \ldots \oplus E_r.$$

In this situation one has a series of first-order differential operators on E,

$$D_i : \Gamma(E) \xrightarrow{\nabla^E} \Gamma(T^*M \otimes E) \xrightarrow{\mathrm{pr}_{E_i}} \Gamma(E_i), \qquad i = 1, \ldots, r.$$

Often, the solution spaces of the differential equations $D_i\varphi = 0$ have interesting relations to geometry. In particular, this is the case if E is associated to a G-structure on M, i.e., $E := P \times_G V$ for a subbundle P of the frame bundle. A natural spitting of $T^*M \otimes E$ is then given by the decomposition of the G-representation on the fibre $(\mathbb{R}^n)^* \otimes V$ into irreducible sub-representations. In the next two sections we want to demonstrate this procedure for the case of k-forms and the case of spinors on semi-Riemannian manifolds with the natural covariant derivative induced by the Levi-Civita connection. We will explain how objects of conformal Cartan geometry, in particular ∇^{nor}-parallel sections, are related to solutions of one of the natural differential equations $D_i\varphi = 0$ in each case.

Let us start with the bundle of k-forms on an n-dimensional semi-Riemannian manifold (M, g) equipped with the Levi-Civita connection ∇^g. A splitting of $T^*M \otimes \Lambda^k(T^*M)$ is given by[6]

$$T^*M \otimes \Lambda^k(T^*M) \simeq \Lambda^{k-1}(T^*M) \oplus \Lambda^{k+1}(T^*M) \oplus C^k(M),$$

where the first two bundles are the images of the contraction map $\lrcorner$ and of the wedge product $\wedge$, respectively, and $C^k(M) \subset T^*M \otimes \Lambda^k(T^*M)$ denotes the intersection of the kernels of $\lrcorner$ and $\wedge$. The projections are

$$\begin{aligned} \mathrm{pr}_{\Lambda^{k-1}}(\sigma \otimes \alpha) &= \sigma^\sharp \lrcorner \alpha, \\ \mathrm{pr}_{\Lambda^{k+1}}(\sigma \otimes \alpha) &= \sigma \wedge \alpha, \end{aligned}$$

and $\mathrm{pr}_{C^k(M)}(\sigma \otimes \alpha)$ viewed as a 1-form with values in $\Lambda^k(T^*M)$ is given by

$$\mathrm{pr}_{C^k(M)}(\sigma \otimes \alpha)v = \sigma(v)\alpha - \frac{1}{k+1} v \lrcorner (\sigma \wedge \alpha) - \frac{1}{n-k+1} v^\flat \wedge (\sigma^\sharp \lrcorner \alpha).$$

The projection of $\nabla^g\alpha$ onto the summand $\Lambda^{k-1}(T^*M)$ yields the codifferential $-d^*\alpha$, the projection of $\nabla^g\alpha$ onto $\Lambda^{k+1}(T^*M)$ is just the differential $d\alpha$. The projection onto the third summand $C^k(M)$ gives an interesting new operator, the so-called *twistor operator* $T : \Gamma(\Lambda^k(T^*M)) \to \Gamma(C^k(M))$, which is described by

$$T(\alpha)(X) := \left(\mathrm{pr}_{C^k(M)} \circ \nabla^g\right)(X) = \nabla^g_X\alpha - \frac{1}{k+1} X \lrcorner d\alpha + \frac{1}{n-k+1} X^\flat \wedge d^*\alpha.$$

The differential equation $T\alpha = 0$ is called the *twistor equation* on k-forms. First, let us take a short look at the twistor equation on 1-forms. A 1-form α is a solution of the twistor equation iff

$$\nabla^g_X\alpha - \frac{1}{2} X \lrcorner d\alpha + \frac{1}{n} X^\flat \wedge d^*\alpha = 0 \quad \forall\, X \in \mathfrak{X}(M). \tag{2.5.1}$$

It is easy to check that the dual vector field of a 1-form α is a conformal Killing field exactly if (2.5.1) is satisfied. This motivates the following name for the solutions of the twistor equation.

[6]This splitting corresponds to the splitting of the $\mathrm{O}(p, q)$-representation $(\mathbb{R}^n)^* \otimes \Lambda^k(\mathbb{R}^n)^*$ into subrepresentations.

Definition 2.5.1. Let $\alpha \in \Omega^k(M)$ be a k-form on a semi-Riemannian manifold (M,g). α is called a conformal Killing form, if

$$\nabla^g_X \alpha - \frac{1}{k+1} X \lrcorner\, d\alpha + \frac{1}{n-k+1} X^\flat \wedge d^*\alpha = 0 \quad \forall\, X \in \mathfrak{X}(M).$$

Next we want to discuss the relation between conformal Killing forms and conformal holonomy. We now consider a conformal manifold (M,c) of signature (p,q) and study parallel sections in the tractor bundle with $(k+1)$-forms as fibres. For this aim, let

$$\rho_{k+1} : \mathrm{O}(p+1,q+1) \longrightarrow \Lambda^{k+1}(\mathbb{R}^{p+1,q+1})^*$$

be the representation of the pseudo-orthogonal group $G = \mathrm{O}(p+1,q+1)$ on the space of $(k+1)$-forms $\Lambda^{k+1}(\mathbb{R}^{p+1,q+1})^*$, induced by the standard representation. We denote the associated tractor bundle by

$$\Lambda^{k+1}_{\mathcal{T}}(M) := \mathcal{P}^1 \times_B \Lambda^{k+1}(\mathbb{R}^{p+1,q+1})^*$$

and call it the *tractor* $(k+1)$*-form bundle.* The scalar product $\langle\cdot,\cdot\rangle_{p+1,q+1}$ on $\mathbb{R}^{p+1,q+1}$ defines an $\mathrm{O}(p+1,q+1)$-invariant scalar product on $\Lambda^{k+1}(\mathbb{R}^{p+1,q+1})^*$, and therefore, a bundle metric on $\Lambda^{k+1}_{\mathcal{T}}(M)$. The normal conformal Cartan connection gives us a metric covariant derivative

$$\nabla^{nor} : \Gamma(\Lambda^{k+1}_{\mathcal{T}}(M)) \longrightarrow \Gamma(T^*M \otimes \Lambda^{k+1}_{\mathcal{T}}(M)).$$

Now, we can proceed in the same way as we described in Section 2.2.4 for the standard tractor bundle. We fix a metric g in the conformal class and reduce $\mathcal{P}^1$ to the frame bundle $\mathcal{P}^g$. If we restrict ρ_{k+1} to the orthogonal group $\mathrm{O}(p,q)$, we obtain a splitting of the $(k+1)$-forms into $\mathrm{O}(p,q)$-invariant subspaces

$$\Lambda^{k+1}(\mathbb{R}^{p+1,q+1})^* \simeq \Lambda^{k}(\mathbb{R}^{p,q})^* \oplus \Lambda^{k+1}(\mathbb{R}^{p,q})^* \oplus \Lambda^{k-1}(\mathbb{R}^{p,q})^* \oplus \Lambda^{k}(\mathbb{R}^{p,q})^*,$$

where this splitting is given by decomposing a $(k+1)$-form $\alpha \in \Lambda^{k+1}(\mathbb{R}^{p+1,q+1})^*$ in the form

$$\alpha = f_0^\flat \wedge \alpha_0 + \alpha_1 + f_0^\flat \wedge f_{n+1}^\flat \wedge \alpha_2 + f_{n+1}^\flat \wedge \alpha_3,$$

with forms α_0, α_1, α_2 and α_3 on $\mathbb{R}^{p,q}$ of degree k, $k+1$, $k-1$ and k, respectively. Hence, fixing a metric $g \in c$, we obtain a splitting

$$\begin{aligned} \Lambda^{k+1}_{\mathcal{T}}(M) &\overset{g}{\simeq} \Lambda^{k}(M) \oplus \Lambda^{k+1}(M) \oplus \Lambda^{k-1}(M) \oplus \Lambda^{k}(M) \\ \alpha &\overset{g}{\simeq} \quad (\alpha_0, \alpha_1, \alpha_2, \alpha_3\,). \end{aligned}$$

With formula (2.2.10) of Section 2.2.4 one obtains for the normal connection ∇^{nor} on $\Lambda^{k+1}_{\mathcal{T}}(M)$ in this splitting,

$$\nabla^{nor}_X \alpha \overset{g}{=} \begin{pmatrix} \nabla^g & -X \lrcorner & X^\flat \wedge & 0 \\ \mathsf{P}(X)^\flat \wedge & \nabla^g_X & 0 & X^\flat \wedge \\ -\mathsf{P}(X) \lrcorner & 0 & \nabla^g_X & X \lrcorner \\ 0 & -\mathsf{P}(X) \lrcorner & -\mathsf{P}(X)^\flat \wedge & \nabla^g_X \end{pmatrix} \begin{pmatrix} \alpha_0 \\ \alpha_1 \\ \alpha_2 \\ \alpha_3 \end{pmatrix}.$$

We now consider $(k+1)$-forms $\alpha \in \Lambda_T^{k+1}(M)$ which are ∇^{nor}-parallel, i.e., which satisfy $\nabla^{nor}\alpha = 0$. We note here that F. Leitner used such ∇^{nor}-parallel forms in order to prove the de Rham Splitting Theorem (Theorem 2.3.2) (cf. [L05a] and [L07]). A closer look at the equation $\nabla^{nor}\alpha = 0$ shows that the forms α_1, α_2 and α_3 in the metric representation of α are uniquely determined by α_0, and moreover, that $\alpha \in \Lambda_T^{k+1}(M)$ is ∇^{nor}-parallel if and only if α_0 satisfies the following equations:

$$0 = \nabla^g_X \alpha_0 - \frac{1}{k+1} X \lrcorner\, d\alpha_0 + \frac{1}{n-k+1} X^\flat \wedge d^*\alpha_0\,, \tag{2.5.2}$$

$$0 = \mathsf{P}(X)^\flat \wedge \alpha_0 + \frac{1}{k+1}\nabla^g_X d\alpha_0 + X^\flat \wedge \Box_k \alpha_0\,, \tag{2.5.3}$$

$$0 = -\mathsf{P}(X) \lrcorner\, \alpha_0 + \frac{1}{n-k+1}\nabla^g_X d^*\alpha + X \lrcorner\, \Box_k \alpha_0\,, \tag{2.5.4}$$

$$0 = -\frac{1}{k+1}\mathsf{P}(X) \lrcorner\, d\alpha_0 - \frac{1}{n-k+1}\mathsf{P}(X)^b \wedge d^*\alpha_0 + \nabla^g \Box_k \alpha_0\,, \tag{2.5.5}$$

where $\Box_k$ is the operator

$$\Box_k := \frac{1}{n-2k}\Big(- \frac{\mathrm{scal}^g}{2(n-1)}\mathrm{Id} + (\nabla^g)^*\nabla^g \Big) \qquad \text{for } n \neq 2k$$

and

$$\Box_{n/2} := \frac{1}{n}\Big[\frac{1}{k+1}(d^*d - dd^*) + \sum_{i=1}^{n} \varepsilon_i \big(s_i^\flat \wedge (\mathsf{P}(s_i) \lrcorner\, \cdot) - s_i \lrcorner\, (\mathsf{P}(s_i)^\flat \wedge \cdot)\big)\Big].$$

The equation (2.5.2) tells us that the k-form α_0 is a conformal Killing form on (M,g). Therefore, we define the following subclass of conformal Killing forms on a semi-Riemannian manifold.

Definition 2.5.2. A conformal Killing k-form α_0 on a semi-Riemannian manifold (M,g) is called a normal conformal Killing form, if it satisfies the additional conditions (2.5.3), (2.5.4) and (2.5.5).

This gives us the following proposition.

Proposition 2.5.1. *Let (M,g) be a semi-Riemannian manifold and $c = [g]$ the conformal class of g. Then there is a 1-1-correspondence between the space*

$$\{\alpha \in \Gamma(\Lambda_T^{k+1}(M)) \mid \nabla^{nor}\alpha = 0\}$$

of all parallel tractor $(k+1)$-forms on (M,c) and the space of all normal conformal Killing k-forms on (M,g).

This shows that special kinds of conformal Killing forms, namely the normal ones, can be studied using conformal holonomy theory. There exists a normal conformal Killing form on (M,g) if and only if the action of the conformal holonomy group $\mathrm{Hol}(M,c) \subset \mathrm{O}(p+1,q+1)$ on $\Lambda^{k+1}(\mathbb{R}^{p+1,q+1})^*$ has a fixed point.

2.6 Conformal holonomy and conformal Killing spinors

In this section we deal with the twistor equation on spinors and describe the relation between the solutions of the twistor equation on spinors (called conformal Killing spinors) and the conformal holonomy group. First we recall some basic notions from spin geometry. Then we define conformal Killing spinors and describe some of their basic properties. Afterwards we show that conformal Killing spinors correspond to parallel sections in the spin tractor bundle for the normal tractor derivative. Therefore, these spinors are directly related to conformal holonomy groups. For a detailed introduction to spin geometry of Riemannian and pseudo-Riemannian manifolds we refer to [Ba81], [ML89], [Fr00] or [Gi09].

Let $(M^{p,q}, g)$ be a space- and time-oriented semi-Riemannian manifold of dimension $n = p + q \geq 2$. We denote by

$$\lambda : \mathrm{Spin}(p,q) \longrightarrow \mathrm{SO}(p,q)$$

the 2-fold covering of the special orthogonal group by the spin group. In this section, $\mathcal{P}^g$ is the bundle of all *oriented* pseudo-orthonormal frames with respect to the given space- and time-orientation. A *spin structure* of (M, g) is a $\mathrm{Spin}^0(p,q)$-principal fibre bundle $\mathcal{Q}^g$ over M with a smooth map $f^g : \mathcal{Q}^g \to \mathcal{P}^g$ which respects the bundle projections and the group actions, i.e.,

$$\begin{aligned} &\pi_{\mathcal{P}^g} \circ f^g = \pi_{\mathcal{Q}^g}, \\ &f^g(u \cdot A) = f^g(u) \cdot \lambda(A) \qquad \forall\, u \in \mathcal{Q}^g,\ A \in \mathrm{Spin}^0(p,q). \end{aligned}$$

It is known that a space- and time-oriented semi-Riemannian manifold (M, g) admits a spin structure if $w_2(M) = 0$, where w_2 is the second Stiefel-Whitney class. In case $w_2(M) = 0$, the number of non-equivalent spin structures coincides with the number of elements in $H^1(M, \mathbb{Z}_2)$. A space- and time-oriented semi-Riemannian manifold $(M^{p,q}, g)$ with a fixed spin structure $(\mathcal{Q}^g, f^g)$ is called a *spin manifold*.

There is a fundamental representation of the spin group, which does not come from the orthogonal group, the *spin representation*

$$\kappa : \mathrm{Spin}(p,q) \longrightarrow \mathrm{GL}(\Delta_{p,q}).$$

The representation space $\Delta_{p,q}$ is a complex vector space of dimension $2^{[n/2]}$. $\Delta_{p,q}$ is an irreducible $\mathrm{Spin}(p,q)$-module, if n is odd, and splits into two irreducible submodules $\Delta_{p,q} = \Delta^+_{p,q} \oplus \Delta^-_{p,q}$, if n is even. Furthermore, there is a $\mathrm{Spin}(p,q)$-equivariant multiplication of vectors from $\mathbb{R}^{p,q}$ with elements of $\Delta_{p,q}$, called the Clifford multiplication. Moreover, there is a $\mathrm{Spin}^0(p,q)$-invariant hermitian form $\langle \cdot, \cdot \rangle_{\Delta_{p,q}}$ on $\Delta_{p,q}$, which is positive definite, if $p = 0, n$, and indefinite, if $1 \leq p \leq n$. An explicit realization of the spin representation can be found for example in [Ba81] or [Fr00]. Using the invariance properties of the Clifford multiplication and of the hermitian form $\langle \cdot, \cdot \rangle_{\Delta_{p,q}}$, we obtain the following four objects on a spin manifold $(M^{p,q}, g)$:

- the *spinor bundle* $\mathcal{S}^g := \mathcal{Q}^g \times_{\mathrm{Spin}^0(p,q)} \Delta_{p,q}$, which is a complex vector bundle of rank $2^{[n/2]}$,
- the *Clifford multiplication* $\mu^g : \begin{array}{ccc} TM \otimes \mathcal{S}^g & \longrightarrow & \mathcal{S}^g \\ (X,\varphi) & \longmapsto & X \cdot \varphi \end{array}$,
- the *spinor derivative* $\nabla^{\mathcal{S}^g} : \Gamma(\mathcal{S}^g) \to \Gamma(T^*M \otimes \mathcal{S}^g)$, which is the covariant derivative induced by the lift of the Levi-Civita connection to $\mathcal{Q}^g$, and
- a *hermitian bundle metric* $\langle \cdot, \cdot \rangle_{\mathcal{S}^g}$ on $\mathcal{S}^g$ defined by $\langle \cdot, \cdot \rangle_{\Delta_{p,q}}$, which is positive definite, if g is Riemannian, and indefinite, if g is pseudo-Riemannian.

Thereby, the following rules are satisfied:

$$\big(X \cdot Y + Y \cdot X\big) \cdot \varphi = -2\, g(X,Y)\, \varphi, \tag{2.6.1}$$

$$\langle X \cdot \varphi, \psi \rangle_{\mathcal{S}^g} = (-1)^{p+1} \langle \varphi, X \cdot \psi \rangle_{\mathcal{S}^g}, \tag{2.6.2}$$

$$X \langle \varphi, \psi \rangle_{\mathcal{S}^g} = \langle \nabla^{\mathcal{S}^g}_X \varphi, \psi \rangle_{\mathcal{S}^g} + \langle \varphi, \nabla^{\mathcal{S}^g}_X \psi \rangle_{\mathcal{S}^g}, \tag{2.6.3}$$

$$\nabla^{\mathcal{S}^g}_X (Y \cdot \varphi) = \nabla^g_X Y \cdot \varphi + Y \cdot \nabla^{\mathcal{S}^g}_X \varphi, \tag{2.6.4}$$

where $X, Y \in \mathfrak{X}(M)$ and $\varphi, \psi \in \Gamma(\mathcal{S}^g)$. Furthermore, any spinor field $\varphi \in \Gamma(\mathcal{S}^g)$ defines a vector field $V_\varphi \in \mathfrak{X}(M)$, called the *Dirac current of* φ, by

$$g(V_\varphi, X) = i^{p+1} \langle X \cdot \varphi, \varphi \rangle_{\mathcal{S}^g} \qquad \forall\, X \in \mathfrak{X}(M). \tag{2.6.5}$$

Obviously, we can extend the Clifford multiplication to a multiplication of k-forms with spinors. For a k-form ω and a spinor φ we set

$$\mu^g(\omega, \varphi) := \omega \cdot \varphi := \sum_{1 \le i_1 < \ldots < i_k \le n} \varepsilon_{i_1} \cdot \ldots \cdot \varepsilon_{i_k}\, \omega(s_{i_1}, \ldots, s_{i_k})\, s_{i_1} \cdot \ldots \cdot s_{i_k} \cdot \varphi,$$

where $(s_1, \ldots, s_n)$ is an orthonormal basis for g and $\varepsilon_i := g(s_i, s_i) = \pm 1$. Then, to any spinor field we can associate a series of k-forms ω^k_φ, where $\omega^1_\varphi = V^\flat_\varphi$, by

$$g(\omega^k_\varphi, \sigma^k) := i^{(p+1)k + \frac{k(k-1)}{2}} \langle \sigma^k \cdot \varphi, \varphi \rangle_{\mathcal{S}^g} \qquad \forall\, \sigma^k \in \Omega^k(M), \tag{2.6.6}$$

where g denotes the induced metric on the space of k-forms.

Now, we define first-order differential operators on the spinor bundle $\mathcal{S}^g$ as described in Section 2.5. We split $T^*M \otimes \mathcal{S}^g$ into a sum of subbundles[7] using the Clifford multiplication $\mu^g : T^*M \otimes \mathcal{S}^g \to \mathcal{S}^g$,

$$T^*M \otimes \mathcal{S}^g \simeq \mathrm{Im}\, \mu^g \oplus \ker \mu^g = \mathcal{S}^g \oplus \ker \mu^g.$$

[7] This decomposition corresponds to the decomposition of the fibre $(\mathbb{R}^n)^* \otimes \Delta_{p,q}$ of $T^*M \otimes \mathcal{S}^g$ into $\mathrm{Spin}(p,q)$-invariant submodules.

We obtain two differential operators of first order by composing the spinor derivative ∇^{S^g} with the projections onto each of these subbundles, namely the *Dirac operator* $D(g)$,

$$D(g) : \Gamma(\mathcal{S}^g) \xrightarrow{\nabla^{S^g}} \Gamma(T^*M \otimes \mathcal{S}^g) \simeq \Gamma(\mathcal{S}^g \oplus \ker \mu^g) \xrightarrow{\mathrm{pr}_{\mathcal{S}^g}} \Gamma(\mathcal{S}^g)$$

and the *twistor operator* $P(g)$,

$$P(g) : \Gamma(\mathcal{S}^g) \xrightarrow{\nabla^{S^g}} \Gamma(T^*M \otimes \mathcal{S}^g) \simeq \Gamma(\mathcal{S}^g \oplus \ker \mu^g) \xrightarrow{\mathrm{pr}_{\ker \mu^g}} \Gamma(\ker \mu^g).$$

Locally, these operators are given by

$$\begin{aligned} D(g)\varphi &= \sum_{i=1}^{n} \sigma^i \cdot \nabla^{S^g}_{s_i} \varphi, \\ P(g)\varphi &= \sum_{i=1}^{n} \sigma^i \otimes \left(\nabla^{S^g}_{s_i} \varphi + \frac{1}{n} s_i \cdot D(g)\varphi \right), \end{aligned}$$

where $(s_1, \dots, s_n)$ is a local orthonormal basis and $(\sigma^1, \dots, \sigma^n)$ its dual. Both operators are conformally covariant. More precisely, if $\tilde{g} = e^{2f} g$ is a conformal change of the metric, the Dirac and the twistor operator change by

$$\begin{aligned} D(\tilde{g}) &= e^{-\frac{n+1}{2} f} D(g) e^{\frac{n-1}{2} f}, \\ P(\tilde{g}) &= e^{-\frac{f}{2}} P(g) e^{-\frac{f}{2}}. \end{aligned}$$

A proof of these formulas can be found in [Gi09], Chapter 1.3 and A.2.

The conformal covariance relates the study of solutions of the *Dirac equation* $D(g)\varphi = 0$, called *harmonic spinors*, and the study of solutions of the *twistor equation* $P(g)\varphi = 0$, called *twistor spinors*, to conformal geometry: it is enough to study these equations for an appropriate "simple" metric in the conformal class of g. Note that a spinor field $\varphi \in \Gamma(\mathcal{S}^g)$ is a solution of the twistor equation $P(g)\varphi = 0$ if and only if it satisfies the equation

$$\nabla^{S^g}_X \varphi = -\frac{1}{n} X \cdot D(g)\, \varphi \qquad \text{for all vector fields } X. \tag{2.6.7}$$

A direct calculation shows that the Dirac-current V_φ of a twistor spinor φ is a conformal vector field. For this reason, twistor spinors are often called also *conformal Killing spinors.* We will use this name further on. More generally, one can prove

Proposition 2.6.1. *Let $\varphi \in \Gamma(\mathcal{S}^g)$ be a conformal Killing spinor. Then for any k, the k-form ω^k_φ is a conformal Killing form.*

Studying the Dirac and the twistor equation more closely, it turns out that they are rather different in nature. We will see that in dimension $n \geq 3$, the twistor equation on spinors has the same nature as the twistor equation on k-forms; the solutions are given by parallel sections of a certain covariant derivative on a certain vector bundle. In particular, the dimension of the solution space of the twistor equation is bounded by a constant depending only on the dimension of M, whereas the dimension of the solution space of the Dirac equation can be arbitrarily large (and for indefinite metrics also infinite). To make this more precise, we first note the following integrability condition for conformal Killing spinors, which is the result of differentiating (2.6.7).

Proposition 2.6.2. ([BFGK91], Th.1.3) *Let $\varphi \in \Gamma(S^g)$ be a conformal Killing spinor on an n-dimensional semi-Riemannian spin manifold, $n \geq 3$. Then,*

$$\nabla_X^{S^g} D(g)\varphi = -\frac{n}{2}\,\mathsf{P}^g(X)\cdot\varphi. \tag{2.6.8}$$

Now, let us consider the doubled spinor bundle $E^g = S^g \oplus S^g$ and the covariant derivative ∇^{E^g} on E^g defined by

$$\nabla_X^{E^g} := \begin{pmatrix} \nabla_X^{S^g} & X\cdot \\ +\frac{1}{2}\,\mathsf{P}^g(X)\cdot & \nabla_X^{S^g} \end{pmatrix}. \tag{2.6.9}$$

The curvature of ∇^{E^g} is given by

$$R^{\nabla^E}(X,Y)\begin{pmatrix} \varphi \\ \psi \end{pmatrix} = \frac{1}{2}\begin{pmatrix} \mathsf{W}(X\wedge Y)\cdot\varphi \\ \mathsf{W}(X\wedge Y)\cdot\psi + \mathsf{C}(X,Y)\cdot\varphi \end{pmatrix}. \tag{2.6.10}$$

For a 3-dimensional manifold (M^3,g) the Weyl tensor W vanishes. (M^3,g) is conformally flat iff $\mathsf{C}=0$. For dimension $n \geq 4$, (M^n,g) is conformally flat iff $\mathsf{W}=0$. $\mathsf{C}=0$ follows from this condition. Hence, by formula (2.6.10), a spin manifold (M,g) is conformally flat exactly if the curvature $R^{\nabla^{E^g}}$ vanishes.

Then, a direct calculation using (2.6.7) and (2.6.8) yields

Proposition 2.6.3. *Let (M^n,g) be a semi-Riemannian spin manifold and $\varphi,\psi \in \Gamma(S^g)$. Then:*

1. *If φ is a conformal Killing spinor, then* $\nabla^{E^g}\begin{pmatrix} \varphi \\ \frac{1}{n}D\varphi \end{pmatrix} = 0$.

2. *If* $\nabla^{E^g}\begin{pmatrix} \varphi \\ \psi \end{pmatrix} = 0$, *then φ is a conformal Killing spinor and* $\psi = \frac{1}{n}D\varphi$.

In this way, conformal Killing spinors can be viewed as parallel sections of (E^g,∇^{E^g}). In particular, we obtain

Proposition 2.6.4. 1. *The dimension of the space* $\ker P$ *of conformal Killing spinors is a conformal invariant bounded by*

$$\dim\ker P \leq 2\,\mathrm{rank}\,S = 2^{[\frac{n}{2}]+1} =: d_n.$$

2. *If* $\dim \ker P = d_n$*, then* (M^n, g) *is conformally flat.*
3. *If* (M^n, g) *is simply connected and conformally flat, then* $\dim \ker P = d_n$.

Hence, for example, all simply connected spaces of constant sectional curvature admit the maximal number of linearly independent conformal Killing spinors. For readers who want to read the latter considerations in the language of physicists, we suggest [PR86].

Since conformal Killing spinors correspond to ∇^{E^g}-parallel sections in the doubled spinor bundle E^g, we can use the holonomy group of the covariant derivative ∇^{E^g} to study the existence of conformal Killing spinors. In the following we want to give a reinterpretation of this situation in the framework of conformal Cartan geometry. We will show that the bundle E^g is isomorphic to the spin tractor bundle over the conformal manifold $(M, c = [g])$, and that the covariant derivative ∇^{E^g} is the normal covariant derivative on this tractor bundle. Hence, the holonomy group of ∇^{E^g} is just given by the holonomy group of the conformal structure $c = [g]$ of the metric g in question. Therefore, the conformal holonomy group $\mathrm{Hol}(M, c)$ contains full information about the existence of conformal Killing spinors. Whereas the classification results for conformal Killing spinors on *Riemannian* spin manifolds were obtained by studying the twistor equation for a single metric, in the more involved case of pseudo-Riemannian metrics in recent years the benefit of conformal Cartan geometry became obvious. A geometric characterization of the geometric structures does not seem to work without using these methods.

Before we describe conformal Killing spinors in the setting of conformal Cartan geometry, we will first have a look at the Riemannian situation and sum up the essential result in this case.

Obviously, parallel spinors, i.e., $\varphi \in \Gamma(\mathcal{S}^g)$ with $\nabla^{\mathcal{S}^g}\varphi = 0$, are harmonic as well as conformal Killing spinors. Moreover,

$$\{\text{harmonic spinors}\} \cap \{\text{conformal Killing spinors}\} = \{\text{parallel spinors}\}.$$

Besides parallel spinors, there is another special class of conformal Killing spinors, the Killing spinors: A spinor field $\varphi \in \Gamma(\mathcal{S}^g)$ is called a *Killing spinor*[8] with *Killing number* ν, if

$$\nabla^{\mathcal{S}^g}_X \varphi = \nu X \cdot \varphi \tag{2.6.11}$$

for a complex number $\nu \in \mathbb{C} \setminus \{0\}$. Killing spinors are twistor spinors as well as eigenspinors of $D(g)$, more precisely,

$$\{\varphi \mid D\varphi = \lambda\varphi\} \cap \{\text{conformal Killing spinors}\} = \left\{ \begin{array}{c} \text{Killing spinors} \\ \text{to Killing number } -\frac{\lambda}{n} \end{array} \right\}.$$

[8] In differential geometry this kind of spinor fields are called Killing spinors. In physics the name Killing spinor is used in a broader sense.

This leads to an important relation between Killing spinors and spectral properties of the Dirac operator on Riemannian spin manifolds, which caused an intensive study of Killing spinors in Riemannian geometry. The Dirac operator on a compact Riemannian spin manifold is (formally) self-adjoint and elliptic. Hence its spectrum contains only real eigenvalues of finite multiplicity.

Proposition 2.6.5. ([Fr81]) *Let (M^n, g) be a compact Riemannian spin manifold with scalar curvature* $\mathrm{scal}^g > 0$. *Then, all eigenvalues λ of $D(g)$ satisfy*

$$\lambda^2 \geq \frac{n}{4(n-1)} \min_M \mathrm{scal}^g. \tag{2.6.12}$$

In particular, there are no harmonic spinors. The equality in (2.6.12) *is attained if and only if the eigenspinor φ with $D(g)\varphi = \lambda\varphi$ is a Killing spinor with Killing number $-\frac{\lambda}{n}$.*

Proof. To be brief, let $D := D(g)$, $P := P(g)$, $\mathcal{S} := \mathcal{S}^g$ and $\nabla := \nabla^{\mathcal{S}^g}$. A direct calculation shows $P^*P = \nabla^*\nabla - \frac{1}{n}D^2$ and $D^2 = \nabla^*\nabla + \frac{1}{4}\mathrm{scal}^g$ (Weitzenböck formula). It follows that

$$D^2 = \frac{n}{n-1}\Big(P^*P + \frac{1}{4}\mathrm{scal}^g\Big). \tag{2.6.13}$$

Then, after integration,

$$\begin{aligned}\int_M \langle D^2\varphi, \varphi\rangle_{\mathcal{S}}\, dM_g &= \frac{n}{n-1}\int_M \Big(\langle P^*P\varphi, \varphi\rangle_{\mathcal{S}} + \frac{1}{4}\mathrm{scal}^g\langle\varphi,\varphi\rangle_{\mathcal{S}}\Big)\, \mathrm{dM_g} \\ &\geq \frac{n}{n-1}\Big(\int_M \langle P\varphi, P\varphi\rangle_{\mathcal{S}}\, dM_g + \frac{1}{4}\min_M \mathrm{scal}^g \int_M \langle\varphi,\varphi\rangle_{\mathcal{S}}\, dM_g\Big).\end{aligned}$$

For a spinor field φ with $D\varphi = \lambda\varphi$ it follows that

$$\lambda^2\|\varphi\|^2_{L_2} \geq \frac{n}{n-1}\Big(\|P\varphi\|^2 + \frac{1}{4}\min_M \mathrm{scal}^g\,\|\varphi\|^2_{L_2}\Big) \geq \frac{n}{4(n-1)}\min_M \mathrm{scal}^g\,\|\varphi\|^2_{L_2},$$

where $\|\varphi\|^2_{L_2} := \int_M \langle\varphi,\varphi\rangle_{\mathcal{S}}\, dM_g$. This shows the estimate (2.6.12). If equality in (2.6.12) holds, then φ satisfies $P\varphi = 0$. Therefore, φ is a Killing spinor with real Killing number $-\frac{\lambda}{n}$. Conversely, if φ is a Killing spinor with real Killing number $-\frac{\lambda}{n}$, then $D\varphi = \lambda\varphi$ and $P\varphi = 0$. With equation (2.6.13), applied to φ, it follows that $\mathrm{scal}^g = \frac{4(n-1)}{n}\lambda^2 = \mathrm{const.}$. Hence, we have the equality in (2.6.12). □

A very nice lecture note on the Dirac spectrum in Riemannian geometry is [Gi09]. There one can find a lot of extensions of the previous proposition and many other relations between the Dirac spectrum and the geometry of the spin manifold.

The geometric structures of Riemannian manifolds with Killing spinors were classified between 1981 and 1993. The results are of interest here, since Killing spinors are special solutions of the twistor equation; even more, they play a crucial role in the description of Riemannian manifolds with conformal Killing spinors.

If a *Riemannian* spin manifold (M^n, g) admits a Killing spinor with Killing number ν, then (M^n, g) is an Einstein space with scalar curvature $\mathrm{scal}^g = 4n(n-1)\nu^2$. (cf. [Fr00] or [Gi09]). Hence, ν is real or purely imaginary. The existence of a parallel spinor ($\nu = 0$) implies that the Riemannian manifold is Ricci-flat. Furthermore, the holonomy group of (M^n, g) is special, since the action of its lift in $\mathrm{Spin}(n)$ has a fixed point on Δ_n. Using Berger's classification of holonomy groups of Riemannian manifolds, this leads to

Proposition 2.6.6. ([Wan89]) *Let (M^n, g) be an irreducible Riemannian spin manifold with non-trivial parallel spinors. Then the reduced holonomy group $\mathrm{Hol}^0(M, g)$ of (M, g) is one of the following groups:*

1. $\mathrm{SU}(m)$, *if* $n = 2m \geq 4$, *(Ricci-flat Kähler manifolds)*,
2. $\mathrm{Sp}(k)$ *if* $n = 4k \geq 8$, *(hyperkähler manifolds)*,
3. G_2, *if* $n = 7$, (G_2*-manifolds*),
4. $\mathrm{Spin}(7)$, *if* $n = 8$, ($\mathrm{Spin}(7)$*-manifolds*).

If (M, g) is a Riemannian spin manifold with real Killing spinor ($\nu \in \mathbb{R}^*$), then the cone $C(M) := \mathbb{R}^+ \times M$ with the cone metric

$$g_C := dt^2 + t^2 \frac{\mathrm{scal}^g}{n(n-1)} g$$

is a Ricci-flat space with parallel spinors. From the geometric structure of the cone, given by its holonomy group, one can read off the geometric structure of (M, g). The result is

Proposition 2.6.7. ([Bär93]) *Let (M, g) be a geodesically complete, simply connected Riemannian spin manifold with real Killing spinor. Then (M, g) is isometric to the standard sphere, or the cone $(C(M), g_C)$ is irreducible and (M, g) has one of the special geometric structures of the following list:*

$\dim(M)$	M	$C(M)$	$\mathrm{Hol}(C(M))$
$2m-1$	*Einstein-Sasaki*	*Ricci-flat, Kähler*	$\mathrm{SU}(2m)$
$4k-1$	3-*Sasaki*	*hyperkähler*	$\mathrm{Sp}(k)$
6	*nearly Kähler*	*generic parallel* 3*-form*	G_2
7	*nearly parallel* G_2	*generic parallel* 4*-form*	$\mathrm{Spin}(7)$

Finally, let φ be an imaginary Killing spinor on (M, g) ($\nu \in i\mathbb{R}^*$). Then the function $u_\varphi := \langle \varphi, \varphi \rangle_{S^g}$ is non-constant. A closer look at the foliation of M into the level sets of u_φ gives the following result, which relates imaginary Killing spinors with parallel spinors in one dimension less.

Proposition 2.6.8. ([B89]) *Let (M, g) be an n-dimensional, geodesically complete Riemannian spin manifold. Then, (M, g) admits a Killing spinor with Killing number $i\alpha \in i\mathbb{R}^*$ if and only if it is isometric to a warped product $(N \times \mathbb{R}, e^{-4\alpha t} h \oplus dt^2)$, where (N, h) is an $(n-1)$-dimensional, geodesically complete Riemannian spin manifold with a non-trivial parallel spinor.*

Next, we will present results concerning Riemannian spin manifolds with *conformal* Killing spinors. For details and proofs we refer to [BFGK91] or [Gi09], and the references therein.

Proposition 2.6.9. *Let (M, g) be a Riemannian spin manifold with a non-trivial conformal Killing spinor φ without zeros. Then $\widetilde{g} := \|\varphi\|_{S^g}^{-4}\, g$ is an Einstein metric of scalar curvature $\mathrm{scal}^{\mathrm{g}} \geq 0$. Moreover, if $\mathrm{scal}^{\widetilde{\mathrm{g}}} > 0$, $(M, \widetilde{g})$ admits real Killing spinors, and if $\mathrm{scal}^{\widetilde{\mathrm{g}}} = 0$, $(M, \widetilde{g})$ admits parallel spinors.*

In view of the previous propositions, it remains to study conformal Killing spinors with zeros as the essential task. On a *Riemannian* spin manifold, the set of zeros $zero(\varphi) := \{x \in M \mid \varphi(x) = 0\}$ of a non-trivial conformal Killing spinor φ is very simple, it is a discrete set of points. The following proposition collects crucial results for the case $zero(\varphi) \neq \emptyset$.

Proposition 2.6.10. *Let (M^n, g) be a Riemannian spin manifold carrying a non-trivial conformal Killing spinor φ with non-empty zero-set $zero(\varphi)$.*

1. *If M^n is compact, then $\mathrm{card}(zero(\varphi)) = 1$ and (M^n, g) is conformally equivalent to the standard sphere (S^n, g_{can}).* ([Li88], [KR00]).

3. *For any number $k \in \mathbb{N}$, there are manifolds (M^n, g) of warped product type with $\mathrm{card}\, zero(\varphi) = k$.* ([BFGK91], [Hab92]).

3. *$M^n \setminus zero(\varphi)$ with the metric $\widetilde{g} := \|\varphi\|^{-4} g$ is a Ricci-flat space with parallel spinors.* ([Gi09], A.2., [KR98], [KR00]).

4. *If the Dirac current V_φ of φ does not vanish, then (M^n, g) is conformally flat, i.e., $\mathsf{W} = 0$.* ([KR94]).

   If $V_\varphi = 0$, there are non-conformally flat solutions. (*Explicit constructions are involved, examples can be found in* [KR96] *and* [KR97].)

In pseudo-Riemannian geometry the situation is much more complicated. Good references for results on parallel, Killing and conformal Killing spinors in the pseudo-Riemannian case are [Ka99], [BK99], [Lei03], [Lei07], [ACGL07], [AC09], [Ba99], [L01], [BL04], [L05a] and [L07] and the references therein. The latter two give a good impression of the benefit of Cartan methods. Readers interested in applications to physics should have a look at [DNP86], [ACDS98], [OV07] and [OHMS09].

Now, we will describe conformal Killing spinors from the view point of conformal Cartan geometry. For this aim, we first extend the basics of conformal Cartan geometry, which we described in section 2.2, to the spin case.

Let (M, c) be a time- and space-oriented conformal manifold of signature (p, q) and $n = p+q \geq 3$. In this section we denote by $\mathcal{P}^0$ the bundle of all time- and space-oriented conformal frames. The structure group is the connected component $\mathrm{CO}^0(p, q) := \mathbb{R}^+ \times \mathrm{SO}^0(p, q)$ of the linear conformal group. Let $\mathrm{CSpin}^0(p, q)$ be the conformal spin group

$$\mathrm{CSpin}^0(p, q) := \mathbb{R}^+ \times \mathrm{Spin}^0(p, q),$$

and let $\lambda : \mathrm{CSpin}^0(p, q) \longrightarrow \mathrm{CO}^0(p, q)$ be the 2-fold covering of the linear conformal group given by the identity in the first component and the usual 2-fold covering of the orthogonal group by the spin group in the second component (which we will denote for simplicity also by λ). A *conformal spin structure* of (M, c) is a $\mathrm{CSpin}^0(p, q)$-principal fibre bundle $\mathcal{Q}^0$ over M with a smooth map $f^0 : \mathcal{Q}^0 \longrightarrow \mathcal{P}^0$, which respects the bundle projections and the group actions, i.e.,

$$\begin{aligned} f^0(q \cdot A) &= f^0(q) \cdot \lambda(A) \qquad \forall\, q \in \mathcal{Q}^0,\ A \in \mathrm{CSpin}_0(p, q), \\ \pi_{\mathcal{P}^0} \circ f^0 &= \pi_{\mathcal{Q}^0}. \end{aligned}$$

We call the conformal manifold (M, c) a *spin manifold*, if it admits a conformal spin structure. This is equivalent to the existence of spin structures for any metric $g \in c$. In fact, any conformal spin structure $(\mathcal{Q}^0, f^0)$ of (M, c) induces a metric spin structure $(\mathcal{Q}^g, f^g)$ for the metric $g \in c$ by reduction

$$\mathcal{Q}^g := (f^0)^{-1}(\mathcal{P}^g) \quad \text{and} \quad f^g := f^0|_{\mathcal{Q}^g},$$

where $\mathcal{P}^g \subset \mathcal{P}^0$ denotes the subbundle of time- and space-oriented g-orthonormal frames. On the other hand, since $\mathcal{P}^0 = \mathcal{P}^g \times_{\mathrm{SO}_0(p,q)} \mathrm{CO}^0(p, q)$, any spin structure $(\mathcal{Q}^g, f^g)$ of a metric $g \in c$ gives rise to a conformal spin structure for the conformal class c by enlargement:

$$\mathcal{Q}^0 := \mathcal{Q}^g \times_{\mathrm{Spin}_0(p,q)} \mathrm{CSpin}^0(p, q) \quad \text{and} \quad f^0 := f^g \times \lambda.$$

Hence, due to the orientability assumption, (M, c) is spin if and only if the second Stiefel-Whitney class of M vanishes: $w_2(M) = 0$.

Now, let $G = \mathrm{SO}^0(p + 1, q + 1)$ be the connected component of the pseudo-orthogonal group and let $B \subset G$ be the matrix realization of the stabilizer of the isotropic line p_∞ in the Möbius sphere $Q^{p,q}$ under the action of the Möbius group $\mathrm{PSO}^0(p+1, q+1)$. We consider the spin group $\widetilde{G} := \mathrm{Spin}^0(p+1, q+1)$ and the lifts of B and its subgroups B_0, B_1 in $\widetilde{G}$: $\widetilde{B} := \lambda^{-1}(B)$, $\widetilde{B}_0 = \lambda^{-1}(B_0)$, $\widetilde{B}_1 = \lambda^{-1}(B_1)$. Then $\widetilde{B}/\widetilde{B}_1$ is isomorphic to the conformal spin group $\widetilde{B}_0 := \mathrm{CSpin}^0(p, q)$. For a conformal spin structure $(\mathcal{Q}^0, f^0)$ of $\mathcal{P}^0$ we define

$$\mathcal{Q}^1 := \{H \subset T_q\mathcal{Q}^0 \mid q \in \mathcal{Q}^0 \text{ and } df^0_q(H) \subset T_{f^0(q)}\mathcal{P}^0 \text{ horizontal and torsion free}\}$$

with the $\widetilde{B}$-action

$$H \cdot \widetilde{b} := (df^0_{q \cdot b_0})^{-1}(df^0_q(H) \cdot \lambda(\widetilde{b})) \quad \forall\, H \subset \mathcal{Q}^1 \cap T_q\mathcal{Q}^0,\ \widetilde{b} = \widetilde{b}_0 \cdot \widetilde{b}_1 \in \widetilde{B} = \widetilde{B}_o \cdot \widetilde{B}_1$$

and the natural projections $\pi^1(H) := q$ and $\pi(H) := \pi^0(q)$. This gives us a $\widetilde{B}$-principal bundle $(\mathcal{Q}^1, \pi, M; \widetilde{B})$ over M and a 2-fold covering

$$f^1 : H \in \mathcal{Q}^1 \longmapsto df^0(H) \in \mathcal{P}^1,$$

which commutes with the group actions and the projections π and π^1. We call the pair $(\mathcal{Q}^1, f^1)$ the *first prolongation of the conformal spin structure* $(\mathcal{Q}^0, f^0)$.

Now, let $\omega \in \Omega^1(\mathcal{P}^1, \mathfrak{g})$ be a Cartan connection on $\mathcal{P}^1$. Since λ_* and df^1_H are isomorphisms, the 1-form

$$\widetilde{\omega} := \lambda_*^{-1} \circ \omega \circ df^1 \in \Omega^1(\mathcal{Q}^1, \widetilde{\mathfrak{g}})$$

is a Cartan connection on $\mathcal{Q}^1$ with values in $\widetilde{\mathfrak{g}} := \mathfrak{spin}(p+1, q+1)$. In particular, the normal conformal Cartan connection ω^{nor} on $\mathcal{P}^1$ defines a normal conformal Cartan connection $\widetilde{\omega}^{nor}$ on $\mathcal{Q}^1$. In the same way as we explained in Section 2.2, any representation

$$\widetilde{\rho} : \mathrm{Spin}(p+1, q+1) \longrightarrow \mathrm{GL}(V)$$

of the spin group defines a tractor bundle $E := \mathcal{Q}^1 \times_{\widetilde{B}} V$ with covariant derivative ∇^{nor}, induced by the normal Cartan connection $\widetilde{\omega}^{nor}$.

Next, we fix again a metric g in the conformal class c and reduce $\mathcal{Q}^1$ to the spin bundle $\mathcal{Q}^g$. Then $E \simeq \mathcal{Q}^g \times_{\mathrm{Spin}^0(p,q)} V$. Let F be one of the bundles TM, T^*M or $\mathfrak{so}(TM, g)$. Analogously to (2.2.9), the choice of g allows us to define the map

$$\widetilde{\rho}^g : F \longrightarrow \mathrm{End}(E, E),$$

$$\widetilde{\rho}^g(\Upsilon)\varphi := \left[\widetilde{u}, \widetilde{\rho}_*\left(\lambda_*^{-1}([u]^{-1}\Upsilon)\right)[\widetilde{u}]^{-1}\varphi\right] \in E_x, \tag{2.6.14}$$

where $\Upsilon \in F_x$, $\varphi \in E_x$ and $\widetilde{u} \in \mathcal{Q}^g_x$.

By ∇^g we denote the covariant derivative on E, which is defined by the Levi-Civita connection of g. In the same way as in Proposition 2.2.4 we obtain

Proposition 2.6.11. *Let* $\widetilde{\rho} : \widetilde{G} \longrightarrow \mathrm{GL}(V)$ *be a representation of the spin group* $\widetilde{G} = \mathrm{Spin}^0(p+1, q+1)$ *and* $E = \mathcal{Q}^1 \times_{\widetilde{B}} V$ *the associated tractor bundle. For any metric* g *in the conformal class* c *the normal tractor connection on* E *is given by*

$$\nabla^{nor}_X = \nabla^g_X + \widetilde{\rho}^g(X) - \widetilde{\rho}^g(\mathsf{P}^g(X)). \tag{2.6.15}$$

The curvature endomorphism of ∇^{nor} *is*

$$R^{nor}(X, Y) = \widetilde{\rho}^g(\mathsf{W}^g(X, Y)) - \widetilde{\rho}^g(\mathsf{C}^g(X, Y)). \tag{2.6.16}$$

Let us apply this to the standard spinor representation

$$\kappa : \mathrm{Spin}^0(p+1, q+1) \longrightarrow \Delta_{p+1,q+1}.$$

We denote the corresponding tractor bundle on a conformal spin manifold (M, c) by

$$\mathcal{S}_{\mathcal{T}}(M) := \mathcal{Q}^1 \times_{\tilde{B}} \Delta_{p+1,q+1}$$

and call it *spin tractor bundle* of (M, c). In the same way as for metric spinor bundles we define a Clifford multiplication

$$\mu : \mathcal{T}(M) \times \mathcal{S}_{\mathcal{T}}(M) \longrightarrow \mathcal{S}_{\mathcal{T}}(M)$$

and a bundle metric $\langle \cdot, \cdot \rangle$ on $\mathcal{S}_{\mathcal{T}}(M)$ using the corresponding $Spin^0$-invariant data on the fibers. Then, spinors $\Phi, \Psi \in \Gamma(\mathcal{S}_{\mathcal{T}}(M))$ satisfy the following rules corresponding to (2.6.1)-(2.6.4), where $Z \in \mathfrak{X}(M)$, $X, Y \in \Gamma(\mathcal{T}(M))$ and $\langle \cdot, \cdot \rangle$ denote the bundle metrics on $\mathcal{T}(M)$ and $\mathcal{S}_{\mathcal{T}}(M)$, respectively:

$$\big(X \cdot Y + Y \cdot X\big) \cdot \Phi = -2 \, \langle X, Y \rangle \, \Phi, \tag{2.6.17}$$

$$\langle X \cdot \Phi, \Psi \rangle = (-1)^{p+2} \langle \Phi, X \cdot \Psi \rangle, \tag{2.6.18}$$

$$Z \, \langle \Phi, \Psi \rangle = \langle \nabla^{nor}_Z \Phi, \Psi \rangle + \langle \Phi, \nabla^{nor}_Z \Psi \rangle, \tag{2.6.19}$$

$$\nabla^{nor}_Z (Y \cdot \Phi) = \nabla^{nor}_Z Y \cdot \Phi + Y \cdot \nabla^{nor}_Z \Phi. \tag{2.6.20}$$

In particular, we can assign to any spinor $\Phi \in \Gamma(\mathcal{S}_{\mathcal{T}}(M))$ a series of tractor k-forms $\alpha^k_\Phi \in \Gamma(\Lambda^k_{\mathcal{T}}(M))$ by

$$\langle \alpha^k_\Phi, \sigma^k \rangle := i^{(p+2)k + \frac{k(k-1)}{2}} \, \langle \sigma^k \cdot \Phi, \Phi \rangle \qquad \forall \, \sigma^k \in \Gamma(\Lambda^k_{\mathcal{T}}(M)). \tag{2.6.21}$$

Next we will prove that any metric $g \in c$ defines a canonical isomorphism between the spin tractor bundle $\mathcal{S}_{\mathcal{T}}(M)$ and the doubled metric spin bundle $E^g = S^g \oplus S^g$ in such a way that the normal tractor connection ∇^{nor} corresponds to the covariant derivative ∇^{E^g} defined in (2.6.9). For this purpose, let us denote by $W_\pm$ the following $\mathrm{Spin}(p, q)$-invariant subspaces of the spinor module $\Delta_{p+1,q+1}$,

$$\begin{aligned} W_- &:= \{ v \in \Delta_{p+1,q+1} \mid f_0 \cdot v = 0 \}, \\ W_+ &:= \{ v \in \Delta_{p+1,q+1} \mid f_{n+1} \cdot v = 0 \}. \end{aligned}$$

As $\mathrm{Spin}(p, q)$-representation, W_+ is isomorphic to $\Delta_{p,q}$. If we restrict the spin representation κ to $\mathrm{Spin}(p, q)$, we obtain the decomposition of $\Delta_{p+1,q+1}$ into the sum of two $\mathrm{Spin}(p, q)$-representations $W_+ \simeq \Delta_{p.q}$:

$$\begin{array}{ccccc} \Delta_{p+1,q+1} & \simeq & W_+ \oplus W_+ & & \\ w_1 + f_0 \cdot w_2 & \mapsto & (w_1, w_2), & & w_1, w_2 \in W_+ . \end{array}$$

Therefore,

$$\mathcal{S}_{\mathcal{T}}(M) \;=\; \mathcal{Q}^g \times_{\mathrm{Spin}_0(p,q)} \Delta_{p+1,q+1} \;\simeq\; \mathcal{Q}^g \times_{\mathrm{Spin}_0(p,q)} (W_+ \oplus W_+) \;\simeq\; S^g \oplus S^g = E^g.$$

The inverse of $\lambda_* : \mathfrak{spin}(p+1,q+1) \to \mathfrak{so}(p+1,q+1)$ applied to $x \in \mathfrak{b}_{-1} \simeq \mathbb{R}^{p,q}$, $z \in \mathfrak{b}_1 \simeq (\mathbb{R}^{p,q})^*$ and $(A,a) \in \mathfrak{b}_0 = \mathfrak{co}(p,q)$ is

$$\begin{aligned} \lambda_*^{-1}(x) &= -\frac{1}{2} x \cdot f_{n+1}, \\ \lambda_*^{-1}(z) &= +\frac{1}{2} z \cdot f_0, \\ \lambda_*^{-1}((A,a)) &= \lambda_*^{-1}(A) + \frac{a}{4}(f_0 \cdot f_{n+1} - f_{n+1} \cdot f_0). \end{aligned}$$

Furthermore, if we represent $v \in \Delta_{p+1.q+1}$ in the form $v = w_1 + f_0 \cdot w_2$ with $w_1, w_2 \in W_+$, then

$$\begin{aligned} f_{n+1} \cdot v = f_{n+1} \cdot f_0 \cdot w_2 &= -f_0 \cdot f_{n+1} \cdot w_2 - 2\langle f_0, f_{n+1}\rangle w_2 = -2w_2, \\ f_0 \cdot v = f_0 \cdot w_1 + f_0 \cdot f_0 \cdot w_2 &= f_0 \cdot w_1. \end{aligned}$$

With these formulas we obtain for the action $\kappa^g(\Upsilon)$ on $\mathcal{S}_\mathcal{T}(M) = \mathcal{S}^g \oplus \mathcal{S}^g$ defined in (2.6.14) in our special case

$$\kappa^g(X)\begin{pmatrix} \varphi \\ \psi \end{pmatrix} = \begin{pmatrix} X \cdot \psi \\ 0 \end{pmatrix}, \qquad \kappa^g(\mu)\begin{pmatrix} \varphi \\ \psi \end{pmatrix} = \begin{pmatrix} 0 \\ -\frac{1}{2}\mu \cdot \varphi \end{pmatrix}, \qquad \kappa^g(\tau)\begin{pmatrix} \varphi \\ \psi \end{pmatrix} = \frac{1}{2}\begin{pmatrix} \tau \cdot \varphi \\ \tau \cdot \phi \end{pmatrix},$$

where $\varphi, \psi \in \Gamma(\mathcal{S}^g)$, X is a vector field, μ a 1-form on M and τ a skew-symmetric endomorphism $\tau \in \Gamma(\mathfrak{so}(TM,g)) = \Gamma(\Lambda^2(T^*M))$.

The application of Proposition 2.6.11 to this special case yields

Proposition 2.6.12. *Let g be a metric in the conformal class c. Then the spin tractor bundle $\mathcal{S}_\mathcal{T}(M)$ can be identified with $\mathcal{S}^g \oplus \mathcal{S}^g$, and in this identification the normal tractor derivative on $\mathcal{S}_\mathcal{T}(M)$ is given by*

$$\nabla^{nor}_X \begin{pmatrix} \varphi \\ \psi \end{pmatrix} = \begin{pmatrix} \nabla^{\mathcal{S}^g}_X \varphi + X \cdot \psi \\ \nabla^{\mathcal{S}^g}_X \psi + \frac{1}{2}\mathsf{P}^g(X) \cdot \varphi \end{pmatrix} = \begin{pmatrix} \nabla^{\mathcal{S}^g}_X & X\cdot \\ +\frac{1}{2}\mathsf{P}^g(X)\cdot & \nabla^{\mathcal{S}^g}_X \end{pmatrix} \begin{pmatrix} \varphi \\ \psi \end{pmatrix}.$$

The curvature of ∇^{nor} is

$$R^{nor}(X,Y)\begin{pmatrix} \varphi \\ \psi \end{pmatrix} = \frac{1}{2}\begin{pmatrix} \mathsf{W}^g(X,Y) \cdot \varphi \\ \mathsf{W}^g(X,Y) \cdot \psi + \mathsf{C}^g(X,Y) \cdot \varphi \end{pmatrix}.$$

This closes the circle and shows that conformal Killing spinors in fact coincide with the ∇^{nor}-parallel sections in the spin tractor bundle $\mathcal{S}_\mathcal{T}(M)$. This opens up a conceptual way of describing all geometric structures admitting conformal Killing spinors. First, try to classify the conformal holonomy groups, then select those groups for which the lift in the spin group stabilizes a spinor in $\Delta_{p+1,q+1}$ and, finally, describe the conformal structures with the resulting holonomy groups.

We finish this section with results on conformal Killing spinors for *Lorentzian* spin manifolds, which were obtained by F. Leitner using methods of conformal

Cartan geometry (cf. [L05a] and [L07]). First of all he was able to prove a local classification result for generic conformal Killing spinors. To state this result, let us note that the Dirac current V_φ of any spinor field $\varphi \in \Gamma(\mathcal{S}^g)$ on a Lorentzian manifold (M, g) is time- or lightlike. We call the spinor φ *generic* if it has no zeros, V_φ does not change the causal type and the dual 1-form $V_\varphi^\flat$ has constant rank, where the rank of a 1-form σ is the number

$$\operatorname{rank}(\sigma) := \max\{k \in \mathbb{N}_0 \mid \sigma \wedge (d\sigma)^k \neq 0\}.$$

Proposition 2.6.13. ([L07], Thm 10) *Let (M, g) be a Lorentzian spin manifold of dimension $n \geq 3$ with a generic conformal Killing spinor φ. Then (M, g) is locally conformal equivalent to one of the following spaces:*

1. $(\mathbb{R}, -dt^2) \times (N_1, h_1) \times \cdots \times (N_r, h_r)$, *where (N_j, h_j) are Ricci-flat Kähler, hyper-Kähler, G_2- or $Spin(7)$-manifolds. This is the case if* $\operatorname{rank}(V_\varphi^\flat) = 0$ *and* $g(V_\varphi, V_\varphi) < 0$.
2. *A Brinkmann space with parallel spinor* ($\operatorname{rank}(V_\varphi^\flat) = 0$ *and* $g(V_\varphi, V_\varphi) = 0$).
3. *A Lorentzian Einstein-Sasaki manifold* (n *odd and* $\operatorname{rank}(V_\varphi^\flat) = \frac{n-1}{2}$).
4. *A Fefferman space* (n *even and* $\operatorname{rank}(V_\varphi^\flat) = \frac{n-2}{2}$).
5. $(N_1, h_1) \times (N_2, h_2)$, *where (N_1, h_1) is a Lorentzian Einstein-Sasaki manifold, and (N_2, h_2) is a Riemannian Einstein-Sasaki manifold, a 3-Sasaki-manifold, a nearly Kähler manifold or a Riemannian sphere. Here* $0 < \operatorname{rank}(V_\varphi^\flat) < \frac{n-2}{2}$.

Proof. We will shortly explain the idea of the proof given in [L07]. Let $\Phi \in \Gamma(\mathcal{S}_{\mathcal{T}}(M))$ be the ∇^{nor}-parallel spinor corresponding to φ. Then, with the help of (2.6.19), (2.6.20) and (2.6.21), one shows that $\nabla^{nor}\alpha_\Phi^k = 0$. Since the signature of $\mathcal{T}(M)$ is $(2, n)$, the 2-form α_Φ^2 does not vanish. Hence, there is a non-zero 2-form $\alpha^2 \in \Lambda^2(\mathbb{R}^{2,n})$ in the stabilizer of the conformal holonomy groups $\operatorname{Hol}(M, [g]) \subset O(2, n)$. The proof then relies on the classification of generic 2-forms on $\mathbb{R}^{2,n}$. The link to the geometric structure of (M, g) results from the observation that the α_--part of α_Φ^2 (cf. Chapter 2.5) is just the Dirac current $V_\varphi^\flat$. □

Conformal Cartan methods have also proved very useful in studying the zero set of conformal Killing spinors. Using the normal form classification for tractor 2-forms, F. Leitner obtained the following description of the zero set of a conformal Killing spinor.

Proposition 2.6.14. ([L07], Thm 19) *Let (M, g) be a Lorentzian spin manifold of dimension $n \geq 3$ with a conformal Killing spinor φ such that $zero(\varphi) \neq \emptyset$. Then $zero(\varphi)$ consists either of*

1. *isolated images of lightlike geodesics and, off the zero set, the metric g is locally conformal equivalent to a Brinkmann space with parallel spinor or*

2. *isolated points and, off the singularity set*[9] $sing(\varphi)$, *the metric is locally conformally equivalent to a static monopole* $-ds^2+h$ *admitting a parallel spinor. In a (convex) neighborhood of* $p \in zero(\varphi)$ *the singularity set is equal to the geodesic light cone, which emerges from* p.

We refer also to [L07] for concrete examples of Lorentzian spin manifolds admitting conformal Killing spinors with zeros.

All Lorentzian manifolds which appear in Proposition 2.6.13 admit global solutions of the conformal Killing spinor equation. In the following chapter we will describe more closely the Fefferman spaces and its conformal Killing spinors. This kind of Lorentzian geometry is the only one, where conformal Killing spinors are 'proper', i.e., where they can not be transformed into parallel or Killing spinors by a conformal change of the metric. We will use the global existence of conformal Killing spinors to prove that the conformal holonomy group of Fefferman spaces is contained in $\mathrm{SU}(1, \frac{n}{2})$. Remember, that $\mathrm{SU}(1, \frac{n}{2})$ is the only proper subgroup of $\mathrm{SO}(2, n)$ that can appear as an irreducibly acting conformal holonomy group of a Lorentzian manifold of dimension $n \geq 4$ (cf. Proposition 2.4.2).

2.7 Lorentzian conformal structures with holonomy group $\mathrm{SU}(1, m)$

Fefferman metrics are Lorentzian metrics on S^1-bundles over strictly pseudoconvex CR manifolds. Such a metric was first discovered by Ch. Fefferman in [F76]. Fefferman studied the boundary behaviour of the Bergman kernel of a strictly pseudoconvex domain $\Omega \subset \mathbb{C}^n$ and in connection with this the solution u of the Dirichlet problem for the complex Monge-Ampere equation on Ω,

$$(-1)^n det \begin{pmatrix} u & \partial u/\partial \bar{z}_k \\ \partial u/\partial z_j & \partial^2 u/\partial z_j \partial \bar{z}_k \end{pmatrix} = 1 \text{ on } \Omega, \qquad u = 0 \text{ on } \partial\Omega. \tag{2.7.1}$$

For existence, uniqueness and regularity of the solution cf. [CY77]. Let u be a solution of (2.7.1), and let $H : cl\,\Omega \times \mathbb{C}^* \to \mathbb{R}$ be the function

$$H(z_1, \ldots, z_n, z_0) = |z_0|^{2/(n+1)} u(z_1, \ldots, z_n).$$

Then the tensor field

$$G = \frac{\partial^2 H}{\partial z_\alpha \partial \bar{z}_\beta} dz_\alpha \odot d\bar{z}_\beta$$

on $cl\,\Omega \times \mathbb{C}^*$ is non-degenerate, and the pull-back $g = j^*G$ for $j : \partial\Omega \times S^1 \hookrightarrow cl\,\Omega \times \mathbb{C}^*$ is a Lorentzian metric on $\partial\Omega \times S^1$. If $\gamma = arg(z_0)$ is the coordinate on S^1, the metric g is given by

$$g = -\frac{i}{n+1} j^*(\partial u - \bar{\partial} u) \odot d\gamma + j^*(\frac{\partial^2 u}{\partial z_j \partial \bar{z}_k} dz_j \odot d\bar{z}_k). \tag{2.7.2}$$

[9]For the exact definition of $sing(\varphi)$ we refer to [L07].

Any biholomorphism of $\Omega \subset \mathbb{C}^n$ lifts to a smooth conformal diffeomorphism of $(\partial\Omega \times S^1, g)$. This indicates a close and interesting link between CR geometry and Lorentzian geometry. Of course, a solution u of (2.7.1) and hence the metric (2.7.2) is not explicitly given. Fefferman used a formal approximation procedure to describe the ambient metric G on a neighborhood of $cl\,\Omega$. Some years later, Farris ([Fa86]) and Lee ([Lee86]) gave an explicit intrinsic description of the Fefferman metric g as a Lorentzian metric on the canonical S^1-bundle over the strictly pseudoconvex CR manifold $\partial\Omega$ and extended the construction to abstract CR manifolds. Graham ([Gr87]) characterized the Fefferman metric locally by curvature properties. Other approaches studied the Fefferman construction and its relation to conformal geometry in the framework of Cartan geometry (see [BDS77], [Ku97], [Ca02] and [Ca06]). Inspired by the close link, the Fefferman construction provides between CR geometry and conformal Lorentzian geometry, there is now a growing number of geometric and analytic investigations of Fefferman spaces. We refer for details to [DT06].

In [Lew91], J. Lewandowski studied the twistor equation for spinors on 4-dimensional space-times and found *local* solutions on Fefferman spaces. In the following section we will explain the intrinsic geometric construction of Fefferman spaces in CR geometry. In particular, we will describe Fefferman spin manifolds, which admit *global* solutions of the conformal Killing spinor equation. Details of the proofs can be found in [Ba99]. As a consequence, we will conclude that the conformal holonomy group of any Fefferman spin manifold is contained in $\mathrm{SU}(1, m)$.

2.7.1 CR geometry and Fefferman spaces

First, we explain the necessary notions from CR geometry. Let N^{2m+1} be a smooth manifold of odd dimension $2m + 1$. A *CR structure on* N is a pair (H, J), where

1. $H \subset TN$ is a real $2m$-dimensional subbundle.
2. $J : H \to H$ is an almost complex structure on H, i.e., $J^2 = -\mathrm{Id}$.
3. If $X, Y \in \Gamma(H)$, then $[JX, Y] + [X, JY] \in \Gamma(H)$ and
   $J([JX, Y] + [X, JY]) - [JX, JY] + [X, Y] \equiv 0$ (integrability condition).

A manifold N equipped with a CR structure (H, J) is called a *CR manifold.* Let us look at some examples.

Example 2.7.1 (Real hypersurfaces of complex manifolds). Any real hypersurface $N \subset \widetilde{N}$ of a complex manifold $(\widetilde{N}, \widetilde{J})$ is a CR manifold. The pair (H, J) is given by $H := TN \cap \widetilde{J}(TN)$ and $J := \widetilde{J}_{|H}$.

Example 2.7.2 (The complex Möbius space). Let $L := \{z \in \mathbb{C}^{m+1} \setminus \{0\} \mid \langle z, z\rangle_{1,m} = 0\}$ be the isotropic cone in the complex vector space $\mathbb{C}^{1,m}$ with hermitian form of signature $(1, m)$. The projectivization $\mathbb{P}L$ is diffeomorphic to the sphere $S^{2m-1} \subset \mathbb{C}^m \simeq \{1\} \times \mathbb{C}^m \subset \mathbb{C}^{1,m}$. Therefore, it admits the canonical

CR structure of Example 2.7.1. Note that $\mathbb{P}L$ is the boundary of the complex hyperbolic space $H^m_{\mathbb{C}} = \{z \in \mathbb{C}^{m+1} \mid \langle z, z\rangle_{1,m} = -1\}$.

Example 2.7.3 (Heisenberg manifolds). Let $He(m)$ be the Heisenberg group

$$He(m) := \left\{ \begin{pmatrix} 1 & x^t & z \\ 0 & \mathbb{I}_m & y \\ 0 & 0 & 1 \end{pmatrix} \,\middle|\, x, y \in \mathbb{R}^m,\ z \in \mathbb{R} \right\}.$$

The Lie algebra of $He(m)$ is the vector space of matrices

$$\mathfrak{he}(m) := \left\{ X(v,w,u) := \begin{pmatrix} 0 & v^t & u \\ 0 & 0 & w \\ 0 & 0 & 0 \end{pmatrix} \,\middle|\, v, w \in \mathbb{R}^m,\ u \in \mathbb{R} \right\}$$

with the commutator

$$\big[X(v,w,u), X(\widetilde{v},\widetilde{w},\widetilde{u})\big] = \big(\langle v, \widetilde{w}\rangle_{\mathbb{R}^m} - \langle \widetilde{v}, w\rangle_{\mathbb{R}^m}\big) \cdot X(0,0,1).$$

A CR structure (H,J) on the Heisenberg group $He(m)$ is given by

$$H := \operatorname{span}\{\widetilde{X}(v,w,0) \mid v, w \in \mathbb{R}^m\} \quad \text{and} \quad J\widetilde{X}(v,w,0) = \widetilde{X}(-w,v,0),$$

where $\widetilde{X}(v,w,u)$ is the left-invariant vector field on $He(m)$ defined by $X(v,w,u)$.

A *Heisenberg manifold* is a manifold of the form $He(m)/\Gamma$, where Γ is a discrete subgroup of $He(m)$. $He(m)/\Gamma$ is a CR manifold with the canonical CR structure induced by the just defined pair (H,J).

Example 2.7.4 (Sasaki manifolds). Let (N,g) be an odd-dimensional Riemannian manifold with a Killing vector field ξ, and let $\varphi := -\nabla^g \xi$. (N,g,ξ) is called a *Sasaki manifold* if

$$\begin{aligned} g(\xi,\xi) &= 1, \\ \varphi^2(X) &= -X + g(X,\xi)\xi \qquad \text{and} \\ (\nabla^g_X \varphi)(Y) &= g(X,Y)\xi - g(Y,\xi)X. \end{aligned}$$

A Riemannian manifold (N,g) admits a Sasaki structure if and only if the cone $(CN = \mathbb{R}^+ \times N, g_C = dt^2 + t^2 g)$ is a Kähler manifold. Obviously, $H := \xi^\perp$ and $J := \varphi_{|H}$ is a CR structure on the Sasaki manifold (N,g,ξ).

Now, let (N,H,J) be an *oriented* CR manifold. In order to define Fefferman spaces we fix a contact form $\theta \in \Omega^1(N)$ such that $\theta|_H = 0$. We call such a contact form θ a *pseudohermitian form* on (N,H,J). Let us denote by T the Reeb vector field of θ, i.e., the vector field T uniquely determined by $\theta(T) = 1$ and $T \lrcorner\, d\theta = 0$. Using the contact form θ we define the *Levi form* L_θ,

$$L_\theta(X,Y) := d\theta(X,JY), \qquad X,Y \in \Gamma(H).$$

Due to the integrability condition of (H,J), the Levi form is symmetric. In the following we suppose that the Levi form L_θ is positive definite. In this case, (N,H,J,θ) is called a *strictly pseudoconvex manifold.* The tensor field

$$g_\theta := L_\theta + \theta \odot \theta$$

defines a Riemannian metric on N. There is a special covariant derivative on a strictly pseudoconvex manifold, the *Tanaka-Webster connection*

$$\nabla^W : \Gamma(TN) \to \Gamma(TN^* \otimes TN).$$

∇^W is metric, i.e., $\nabla^W g_\theta = 0$, and its torsion tensor Tor^W is given by

$$\mathrm{Tor}^W(X,Y) = L_\theta(JX,Y)\cdot T \qquad \text{and}$$
$$\mathrm{Tor}^W(T,X) = -\frac{1}{2}([T,X] + J[T,JX])$$

for $X,Y \in \Gamma(H)$. These two conditions determine ∇^W uniquely. In particular, the Tanaka-Webster connection satisfies $\nabla^W T = 0$ and $\nabla^W(J) = 0$.

Now, let us denote by $T^{10} \subset TN^{\mathbb{C}}$ the eigenspace of the complex-linear extension of J on $H^{\mathbb{C}}$ to the eigenvalue i. Then L_θ extends to a Hermitian form on T^{10} by

$$L_\theta(U,V) := -i\, d\theta(U,\bar{V}), \qquad U,V \in \Gamma(T^{10}).$$

For a complex 2-form $\omega \in \Lambda^2 N^{\mathbb{C}}$ we denote by $\mathrm{trace}_\theta\, \omega$ the θ-trace of ω:

$$\mathrm{trace}_\theta\, \omega := \sum_{\alpha=1}^{m} \omega(Z_\alpha, \bar{Z}_\alpha),$$

where $(Z_1,\dots,Z_m)$ is a unitary basis of (T^{10}, L_θ). Let R^W be the $(4,0)$-curvature tensor of the Tanaka-Webster connection ∇^W on the complexified tangent bundle of N,

$$R^W(X,Y,Z,V) := g_\theta\big(([\nabla^W_X, \nabla^W_Y] - \nabla^W_{[X,Y]})Z, \bar{V}\big), \qquad X,Y,Z,V \in \Gamma(TN^{\mathbb{C}}).$$

Furthermore, we consider the *Tanaka-Webster Ricci curvature*

$$\mathrm{Ric}^W := \mathrm{trace}_\theta^{(3,4)} R^W := \sum_{\alpha=1}^{m} R^W(\cdot,\cdot,Z_\alpha,\bar{Z}_\alpha)$$

and the *Tanaka-Webster scalar curvature*

$$\mathrm{scal}^{\mathrm{W}} := \mathrm{trace}_\theta\, \mathrm{Ric}^{\mathrm{W}}.$$

It is easy to check that $\mathrm{scal}^{\mathrm{W}}$ is a real-valued function on N.

Next, let us consider the canonical line bundle K of the CR manifold (N,H,J),

$$K := \Lambda^{m+1,0}N := \{\omega \in \Lambda^{m+1}N^{\mathbb{C}} \mid V \lrcorner\, \omega = 0 \quad \forall\, V \in \overline{T^{10}}\}.$$

$\mathbb{R}^+$ acts on $K^* = K \setminus \{0\}$ by multiplication. Then,

$$F := K^*/\mathbb{R}^+ \xrightarrow{\ \pi\ } N$$

is the S^1-principal fibre bundle on N associated to K. We call F the canonical S^1-bundle of the CR manifold (N, H, J). There are special Lorentzian metrics on the S^1-bundle F, which are constructed in the following way: We choose a connection form A on F and a constant $c \in \mathbb{R}$ and consider the metric

$$h_{A,c} := \pi^* L_\theta - ic\pi^*\theta \odot A.$$

Since L_θ is positive definite, $h_{A,c}$ is a Lorentzian metric on F. The fibres of F are lightlike submanifolds. Moreover, the fundamental vector fields of the S^1-action on F are lightlike Killing fields. By definition, the metric $h_{A,c}$ depends on the pseudohermitian form θ. We now look for a connection form A and a constant c, such that the conformal class $[h_{A,c}]$ does not depend on θ. First we note that the Tanaka-Webster connection ∇^W defines a connection form A^W on F, which we call Tanaka-Webster connection as well. To define A^W, let us consider a local unitary basis $s = (Z_1, \ldots, Z_m)$ of (T^{10}, L_θ) on $U \subset N$ and the matrix of local 1-forms $\omega_s = (\omega_{\alpha\beta})$ given by

$$\nabla^W Z_\alpha = \sum_\beta \omega_{\alpha\beta} \otimes Z_\beta.$$

Since ∇^W is metric, the trace of the matrix ω_s is purely imaginary. We denote by $(\theta^1, \ldots, \theta^m, \bar\theta^1, \cdots, \bar\theta^m, \theta)$ the dual basis to $(Z_1, \ldots, Z_m, \bar Z_1, \ldots, \bar Z_m, T)$. Then

$$\tau_s := [\theta \wedge \theta^1 \wedge \ldots \wedge \theta^m] : U \subset N \longrightarrow F := K^*/\mathbb{R}^+$$

is a local section in the S^1-bundle F. We define A^W as the connection form on F which satisfies

$$\tau_s^* A^W := -\operatorname{trace} \omega_s : TU \longrightarrow i\mathbb{R}.$$

For the curvature of A^W we get $\Omega^{A^W} = -\operatorname{Ric}^W$. Now we are ready to define a special connection form A and a constant c such that the conformal class $[h_{A,c}]$ does not depend on θ. As is well known, two connection forms on an S^1-bundle differ by a 1-form with imaginary values on the basis of the bundle. For the following modification of the Tanaka-Webster connection

$$A_\theta := A^W - \frac{i}{2(m+1)} \operatorname{scal}^W \cdot \theta \tag{2.7.3}$$

and the constant $c := \frac{4}{m+2}$ one obtains

Proposition 2.7.1. ([Lee86]) *Let (N, H, J, θ) be an oriented strictly pseudoconvex manifold with the canonical S^1-bundle $\pi : F \to N$, and let h_θ be the Lorentzian metric*

$$h_\theta := \pi^* L_\theta - i\frac{4}{m+2}\pi^*\theta \odot A_\theta \tag{2.7.4}$$

on F. If $\widetilde{\theta}$ is another pseudohermitian form on (N, H, J) in the same orientation class as θ, then there exists a positive function f on N such that $\widetilde{\theta} = f \cdot \theta$. The Lorentzian metrics on F satisfy $h_{\widetilde{\theta}} = (f \circ \pi) \cdot h_\theta$. In particular, $(F, [h_\theta])$ is a conformal Lorentzian manifold, associated to the oriented CR-manifold (N, H, J), which does not depend on the choice of θ.

Definition 2.7.1. The Lorentzian manifold (F, h_θ) is called the Fefferman space of the strictly pseudoconvex manifold (N, H, J, θ). h_θ is called the Fefferman metric.

The following proposition explains the local characterization of Fefferman spaces by curvature conditions, which was found by G. Sparling and R. Graham.

Proposition 2.7.2. ([Gr87]), [Sp85]) *Let (N, H, J, θ) be an oriented strictly pseudoconvex manifold, and let (F, h_θ) be its Fefferman space. Then any fundamental vertical vector field ξ on F is lightlike and Killing and satisfies*

$$\begin{aligned} \xi \lrcorner \mathsf{W}^{h_\theta} &= 0, \\ \xi \lrcorner \mathsf{C}^{h_\theta} &= 0, \\ \mathsf{P}^{h_\theta}(\xi, \xi) &= const > 0. \end{aligned}$$

Conversely, let (M, h) be a Lorentzian manifold with a regular lightlike Killing field ξ such that

$$\begin{aligned} \xi \lrcorner \mathsf{W}^{h} &= 0, \\ \xi \lrcorner \mathsf{C}^{h} &= 0, \\ \mathsf{P}^{h}(\xi, \xi) &= const > 0. \end{aligned}$$

Then the leaf space $N := M/\xi$ has the structure of a strictly pseudoconvex manifold and its Fefferman space is locally isometric to (M, h).

Now, we consider again an oriented strictly pseudoconvex manifold (N,H,J,θ) and suppose in addition, that (N, g_θ) has a spin structure. This spin structure defines a square root $\sqrt{K}$ of the canonical line bundle $K = \Lambda^{m+1,0}N$, i.e., a line bundle $\sqrt{K}$ with $\sqrt{K} \otimes \sqrt{K} = K$ (for details of this construction we refer to [Ba99]). Then we proceed in the same way as before. We consider the S^1-bundle $\sqrt{F} := \sqrt{K}^*/\mathbb{R}^+$ over N and the Tanaka-Webster connection A^W on $\sqrt{F}$ defined by ∇^W. Now, $\Omega^{A^W} = -\frac{1}{2}\,\mathrm{Ric}^W$. We modify the Tanaka-Webster connection by

$$A_\theta^{\sqrt{}} := A^W - \frac{i}{4(m+1)}\mathrm{scal}^W \cdot \theta \tag{2.7.5}$$

and consider the Lorentzian metric

$$h_\theta^{\surd} := \pi^* L_\theta - i\,\frac{8}{m+2}\pi^*\theta \odot A_\theta^{\surd} \tag{2.7.6}$$

on $\sqrt{F}$, which we also call a *Fefferman metric.* Here, again, the connection form $A_\theta^{\surd}$ and the constant $\frac{8}{m+2}$ in (2.7.6) are chosen in this way, in order to ensure that the conformal class $[h_\theta^{\surd}]$ does not depend on the choice of θ. Hence, the conformal structure $[h_\theta^{\surd}]$ on $\sqrt{F}$ is an invariant of the CR structure (N,H,J). Now, since (N,g_θ) is spin, the Lorentzian manifold $(\sqrt{F},h_\theta^{\surd})$ is spin as well, where the spin structure is obtained by that of (N,g_θ) in a canonical way. For further purpose we mention the structure of the spinor bundle $\mathcal{S}^{h_\theta^{\surd}}$ of $(\sqrt{F},h_\theta^{\surd})$. Let $(\mathcal{Q}^{g_\theta},f^{g_\theta})$ denote the spin structure of (N,g_θ). The CR structure (H,J) defines a subbundle $\mathcal{Q}_H := (f^{g_\theta})^{-1}(\mathcal{P}_H) \subset \mathcal{Q}^{g_\theta}$, where

$$\mathcal{P}_H := \{u=(X_1,JX_1,\dots,X_m,JX_m,T) \mid u \ g_\theta\text{-orthonormal}\} \subset \mathcal{P}^{g_\theta},$$

and an associated spinor bundle $\mathcal{S}_H := \mathcal{Q}_H \times_{\lambda^{-1}(U(m))} \Delta_{2m}$ on N. Then,

$$\mathcal{S}^{h_\theta^{\surd}} \simeq \pi^*\mathcal{S}_H \oplus \pi^*\mathcal{S}_H.$$

Definition 2.7.2. We call $(\sqrt{F},h_\theta^{\surd})$ with its canonically induced spin structure the Fefferman spin space of the strictly pseudoconvex spin manifold (N,H,J,θ).

With this modification of the topological type of the canonical S^1-bundle, using the spin structure, we can give a spinorial characterization of Fefferman spaces. First, let us note the following structure of the spinor bundle $\mathcal{S}_H$.

Proposition 2.7.3. ([Ba99]) *Let (N,H,J,θ) be a strictly pseudoconvex spin manifold, and let $\mathcal{S}_H$ be the spinor bundle with respect to (H,J). Then,*

1. *$\mathcal{S}_H$ decomposes into a sum of subbundles*

$$\mathcal{S}_H = \bigoplus_{r=0}^{m} \mathcal{S}_{(-m+2r)i},$$

where $\mathcal{S}_{(-m+2r)i}$ are the eigenspaces of the action of the 2-form $d\theta$ on $\mathcal{S}_H$. Thereby,

$$d\theta\cdot|_{\mathcal{S}_{(-m+2r)i}} = (-m+2r)\cdot \mathrm{Id}_{\mathcal{S}_{(-m+2r)i}} \quad \textit{and} \quad \dim \mathcal{S}_{(-m+2r)i} = \binom{m}{r}.$$

2. *The lifted bundles $\pi^*\mathcal{S}_{\pm mi}$ are trivial line bundles on $\sqrt{F}$ with global sections $\psi_\epsilon \in \Gamma(\pi^*\mathcal{S}_{\epsilon mi})$, $\epsilon=\pm 1$.*

Using the global sections ψ_ϵ in the line bundles $\pi^*\mathcal{S}_{\pm mi}$, we obtain a 2-parametric family of conformal Killing spinors on $(\sqrt{F},h_\theta^{\surd})$. Furthermore, using Proposition 2.7.2 we can give a characterization of Fefferman spaces by means of conformal Killing spinors.

Proposition 2.7.4. ([Ba99], [BL04]) *Let (N, H, J, θ) be a strictly pseudoconvex spin manifold with the Fefferman spin space $(\sqrt{F}, h_\theta^{\surd})$. Then the spinor fields $\phi_\epsilon := (\psi_\epsilon, 0) \in \Gamma(\mathcal{S}^{h_\theta^{\surd}}) = \Gamma(\pi^*\mathcal{S}_H \oplus \pi^*\mathcal{S}_H)$, $\epsilon = \pm 1$, are conformal Killing spinors, such that*

a) *the Dirac current V_{ϕ_ϵ} is a regular lightlike Killing field,*

b) $\nabla^{\mathcal{S}^{h_\theta^{\surd}}}_{V_{\phi_\epsilon}} \phi_\epsilon = i\, c\, \phi_\epsilon$, *where $c \in \mathbb{R}\backslash\{0\}$.*

Conversely, if (M, h) is an even-dimensional Lorentzian spin manifold with a conformal Killing spinor ϕ, such that

a) *the Dirac current V_ϕ is a regular lightlike Killing field,*

b) $\nabla^{\mathcal{S}^h}_{V_\phi} \phi = i\, c\, \phi$, *where $c \in \mathbb{R}\backslash\{0\}$,*

then there exists a strictly pseudoconvex spin manifold (N, H, J, θ), such that its Fefferman spin space is locally isometric to (M, h).

2.7.2 Conformal holonomy of Fefferman spaces

Using the results of the previous section, we will now show that the conformal holonomy group of the Fefferman spin manifold $(\sqrt{F}, [h_\theta^{\surd}])$ is contained in $\mathrm{SU}(1, m+1) \subset \mathrm{SO}(2, 2m+2)$. Remember that $(\sqrt{F}, h_\theta^{\surd})$ is a Lorentzian manifold of signature $(1, 2m+1)$. In the first step we show that

$$\mathrm{Hol}(\sqrt{F}, [h_\theta^{\surd}]) \subset \mathrm{U}(1, m+1).$$

For this purpose, we will define an isometric almost complex structure on the standard tractor bundle $\mathcal{T}(\sqrt{F})$, which is parallel with respect to the normal tractor connection ∇^{nor}. The further reduction to $\mathrm{SU}(1, m+1)$ is then a consequence of the existence of the two global conformal Killing spinors ϕ_ϵ.

First, let us consider once more the S^1-bundle $\pi : \sqrt{F} \to N$ associated with the strictly pseudoconvex spin manifold (N, H, J, θ). We denote by $Z^* \in \mathfrak{X}(\sqrt{F})$ the horizontal lift of a vector field Z on N with respect to the connection form $A_\theta^{\surd}$. Furthermore, we fix a vertical fundamental vector field ξ on $\sqrt{F}$ by the condition $h_\theta^{\surd}(T^*, \xi) = 1$, where T is the Reeb vector field of θ. Remember that by definition, ξ and T^* are lightlike. Then, any vector field on $\sqrt{F}$ is of the form

$$\gamma T^* + \delta \xi + X^*,$$

where γ, δ are functions on $\sqrt{F}$ and $X \in \Gamma(H)$.

Now, let $c = [h_\theta^{\surd}]$ be the conformal class of $h_\theta^{\surd}$, and let $\mathcal{T}(\sqrt{F})$ be the standard tractor bundle of the Fefferman spin manifold $(\sqrt{F}, c)$. As we saw in Section 2.2.4, the choice of a metric $h_\theta \in c$ gives rise to a splitting of $\mathcal{T}(\sqrt{F})$ into

$$\mathcal{T}(\sqrt{F}) \stackrel{h_\theta}{\simeq} \underline{\mathbb{R}} \oplus T\sqrt{F} \oplus \underline{\mathbb{R}}.$$

Using this splitting, we define an almost complex structure $J^{\mathcal{T}}$ on $\mathcal{T}(\sqrt{F})$ by

$$J^{\mathcal{T}}\begin{pmatrix} \alpha \\ \gamma T^* + \delta\xi + X^* \\ \beta \end{pmatrix} := \begin{pmatrix} -\frac{\delta}{2} \\ 2\alpha\xi + \frac{\beta}{2}T^* + 2\nabla^{h_\theta}_{X^*}\xi \\ -2\gamma \end{pmatrix}, \tag{2.7.7}$$

where α, β, γ, δ are functions on $\sqrt{F}$, and T^*, ξ, X^* are vector fields on $\sqrt{F}$ defined as above. Then, by a direct but lengthy calculation one obtains the following proposition. For a more conceptional proof see [L05b], [L07] .

Proposition 2.7.5. *The endomorphism $J^{\mathcal{T}} : \mathcal{T}(\sqrt{F}) \longrightarrow \mathcal{T}(\sqrt{F})$ satisfies*

1. $(J^{\mathcal{T}})^2 = -\mathrm{Id}_{\mathcal{T}(\sqrt{F})}$,
2. $\langle J^{\mathcal{T}}\phi_1, J^{\mathcal{T}}\phi_2\rangle_{\mathcal{T}} = \langle\phi_1,\phi_2\rangle_{\mathcal{T}}$ *for all* $\phi_1,\phi_2 \in \mathcal{T}(\sqrt{F})$,
3. $\nabla^{nor}(J^{\mathcal{T}}) = 0$.

According to Proposition 2.1.5, the parallel almost complex structure $J^{\mathcal{T}}$ defines a hermitian almost complex structure J_0 on the fibre $\mathbb{R}^{2,2m+2}$ of $\mathcal{T}(\sqrt{F})$ which is invariant under the action of the conformal holonomy group $\mathrm{Hol}(\sqrt{F},[h_\theta^{\sqrt{}}])$. This means that

$$\mathrm{Hol}(\sqrt{F},[h_\theta^{\sqrt{}}]) \subset \mathrm{U}(1,m+1) \subset \mathrm{SO}(2,2m+2).$$

Now, in order to reduce the situation further to $\mathrm{SU}(1,m+1)$, we use the following algebraic lemma.

Lemma 2.7.1. *Let $A \in \mathrm{U}(1,m+1) \subset \mathrm{SO}(2,2m+2)$, and let $\widetilde{A} \in \lambda^{-1}(A) \subset \mathrm{Spin}(2,2m+2)$ be one of the two elements of the spin group which cover A. If there are two linearly independent spinors in the spinor module $\Delta_{2,2m+2}$, which are fixed by the action of $\widetilde{A}$, then $A \in \mathrm{SU}(1,m+1)$.*

Now, according to Proposition 2.7.4, there exist at least two linearly independent conformal Killing spinors on the Fefferman spin manifold $(\sqrt{F},h_\theta^{\sqrt{}})$. Hence, by Proposition 2.6.12, there are two linearly independent ∇^{nor}-parallel sections in the spin tractor bundle $\mathcal{S}_{\mathcal{T}}(\sqrt{F})$. We use again Proposition 2.1.5, which tells us that the action of the holonomy group $\mathrm{Hol}(\mathcal{S}_{\mathcal{T}}(\sqrt{F}),\nabla^{nor}) \subset \mathrm{Spin}(2,2m+2)$ on the fibre $\Delta_{2,2m+2}$ of $\mathcal{S}_{\mathcal{T}}(\sqrt{F})$ has two linearly independent fixed elements. Since we already know that

$$\lambda\left(\,\mathrm{Hol}(\mathcal{S}_{\mathcal{T}}(\sqrt{F}),\nabla^{nor})\,\right) = \mathrm{Hol}(\sqrt{F},h_\theta^{\sqrt{}}) \subset \mathrm{U}(1,m+1),$$

Lemma 2.7.1 yields

Proposition 2.7.6. *The conformal holonomy group of a Fefferman spin manifold of signature $(1,2m+1)$ is special unitary, i.e.,*

$$\mathrm{Hol}(\sqrt{F},[h_\theta^{\sqrt{}}]) \subset \mathrm{SU}(1,m+1) \subset \mathrm{SO}(2,2m+2)\,.$$

Using the local curvature characterization of Fefferman spaces given by R. Graham (cf. Proposition 2.7.2) and further, more detailed investigations in conformal Cartan geometry, one can prove the converse of Proposition 2.7.6. For details we refer to [L07] and [L08].

Proposition 2.7.7. *Let (M, g) be a Lorentzian manifold of even dimension $n \geq 4$, and assume that the reduced conformal holonomy group is special unitary, i.e.,*

$$\mathrm{Hol}^0(M, [g]) \subset \mathrm{SU}(1, n/2) \subset \mathrm{SO}(2, n).$$

Then (M, g) is locally conformal equivalent to a Fefferman spin space.

2.8 Further results

In the previous section we described conformal structures with holonomy groups in $\mathrm{SU}(1, m+1)$. This, in principle, completes the picture in the Riemannian and in the Lorentzian situation (cf. Section 2.4). We finish these lectures with some remarks on results in higher signature.

First, note that the constructions in the previous section also work if the Levi form L_θ is non-degenerate, but not necessarily positive definite. In case the hermitian space (T^{10}, L_θ) has signature (p, q), with $p + q = m$, one obtains in the same way a pseudo-Riemannian manifold $(\sqrt{F}, h_\theta^{\sqrt{}})$ of signature $(2p+1, 2q+1)$ with conformal holonomy group

$$\mathrm{Hol}(\sqrt{F}, [h_\theta^{\sqrt{}}]) \subset \mathrm{SU}(p+1, q+1) \subset \mathrm{SO}(2p+2, 2q+2),$$

and again, any of such a conformal manifold can be locally obtained by this construction.

The next interesting case is that of pseudo-Riemannian manifolds (M, g) of signature $(3, 4m+3)$ with conformal holonomy group in the symplectic group

$$\mathrm{Sp}(1, m+1) \subset \mathrm{SO}(4, 4m+4).$$

Such manifolds were studied in detail by Jesse Alt in [Al08] and [Al09]. The models of such manifolds are S^3-bundles over a *quaternionic contact manifold* equipped with a canonical conformal structure, introduced by O. Biquard, cf. [Bi06].

For the group $G_{2(2)} \subset \mathrm{SO}(4, 3)$ much has been done in recent work by P. Nurowski, T. Leistner, M. Hammerl and K. Sagerschnig. Conformal manifolds with holonomy in $G_{2(2)}$ are 5-dimensional of signature $(3, 2)$ and admit a generic rank 2 distribution. For further studies of this case we refer to [Nu05], [Nu08], [LN09], [Ha09] and [HaS09].

Conformal manifolds with holonomy group in $\mathrm{Spin}(4, 3) \subset \mathrm{SO}(4, 4)$ have been studied by R. Bryant in [Bry06]. Such manifolds are 6-dimensional of signature $(3, 3)$ and admit a generic rank 3 distribution.

Finally, we want to mention that in [Al08], J. Alt proposed a general concept for the classification of conformal holonomy groups $H \subset O(p+1, q+1)$, which act locally transitively on the Möbius sphere $Q^{p,q}$. This approach includes a generalization of the Fefferman construction, which we carried out in detail for the CR case, to more general cases. Thereby, the interpretation of this construction in the general framework of parabolic Cartan geometry is essential. We refer for this general approach to Fefferman spaces to [Ca06] and [Al08]. Moreover, for all who want to understand the general background of parabolic geometry, which generalizes conformal Cartan geometry, we recommend [CS09] and the forthcoming second volume of this book.

Bibliography

[AGMOO00] O. Aharony, S. Gubser, J. Maldacena, H. Ooguri and Y. Oz. Large N field theories, string theory and gravity. *Phys. Rep.* **323** no. 3-4 (2000), 183–386. `http://arxiv.org/abs/hep-th/9905111v3`

[AC09] D.V. Alekseevsky and V. Cortés. On pseudo-Riemannian manifolds with many Killing spinors. *AIP Conference Proceedings*, volume 1093, March 2009.

[ACDS98] D.V. Alekseevsky, V. Cortés, C. Devchand and U. Semmelmann. Killing spinors are Killing vector fields in Riemannian supergeometry. *J. Geom. Phys.* **26** (1998), 51–78.

[ACGL07] D.V. Alekseevsky, V. Cortés, A. Galaev and T. Leistner. Cones over pseudo-Riemannian manifolds and their holonomy. To appear in *Crelle's Journal.* `http://arxiv.org/abs/0707.3063v2`

[A07] S. Alexakis. The decomposition of global conformal invariants: On a conjecture of Deser and Schwimmer. `http://arxiv.org/abs/0711.1685v1`

[A06a] S. Alexakis. On the decomposition of global conformal invariants I. *Annals of Math.*. `http://arxiv.org/abs/math/0509571v3`

[A06b] S. Alexakis. On the decomposition of global conformal invariants. II. *Advances in Math.* **206** no. 2 (2006), 466–502. `http://arxiv.org/abs/math/0509572v3`

[Al08] J. Alt. *Fefferman constructions in conformal holonomy.* PhD-thesis, Humboldt-Universität, Berlin, 2008.

[Al09] J. Alt. On quaternionic contact Fefferman spaces. Preprint, submitted to Diff. Geom. Appl. 2009.

[A01] M. Anderson, L^2 curvature and volume renormalization of AHE metrics on 4-manifolds. *I. Math. Res. Lett.* **8** no. 1-2 (2001), 171–188. `http://arxiv.org/abs/math/0011051v2`

[Ar06] S. Armstrong. *Holonomy of Cartan connections.* PhD-thesis, Oxford, 2006.

[Ar07] S. Armstrong. Definite signature conformal holonomy: A complete classification. *J. Geom. Phys.* **57** (2007), 2024–2048. `http://arxiv.org/abs/math/0503388v3`

[Bär93] C. Bär. Real Killing spinors and holonomy. *Comm. Math. Phys.* **154** (1993), 509–521.

[BE53] H. Bateman and A. Erdelyi. *Higher transcendental functions*, volume I. 1953.

[Ba81] H. Baum. *Dirac operators and spin structures on pseudo-Riemannian manifolds* (in German), volume 41 of *Teubner-Texte zur Mathematik.* Teubner-Verlag, Stuttgart/Leipzig, 1981.

[B89] H. Baum. Complete Riemannian manifolds with imaginary Killing spinors. *Ann. Glob. Anal. Geom.* **7** (1989), 205–226.

[Ba99] H. Baum. Lorentzian twistor spinors and CR-geometry. *Diff. Geom. and its Appl.* **11** (1999), 69–96.

[Ba09] H. Baum. *Eichfeldtheorie. Eine Einführung in die Differentialgeometrie auf Faserbündeln.* Springer-Verlag, 2009.

[BFGK91] H. Baum, T. Friedrich, R. Grunewald and I. Kath. *Twistors and Killing Spinors on Riemannian Manifolds*, volume 124 of *Teubner-Texte zur Mathematik.* Teubner-Verlag, Stuttgart/Leipzig, 1991.

[BK99] H. Baum and I. Kath. Parallel spinors and holonomy groups on pseudo-Riemannian spin manifolds. *Ann. Glob. Anal. Geom.* **17** (1999), 1–17.

[BL04] H. Baum and F. Leitner. The twistor equation in Lorentzian spin geometry. *Math. Zeitschrift* **247** (2004), 795–812.

[Be93] W. Beckner. Sharp Sobolev inequalities on the sphere and the Moser-Trudinger inequality. *Ann. of Math. (2)* **138** no.1 (1993), 213–242.

[BI93] L. Bérard-Bergery and A. Ikemakhen. On the holonomy of Lorentzian manifolds. In: *Differential Geometry: Geometry in Mathematical Physics and Related Topics*, 27–40. Proc. Symp. Pure Math. 54. Amer. Math. Soc., 1993.

[Be87] A. Besse. *Einstein manifolds*, volume 10 of *Ergebnisse der Mathematik und ihrer Grenzgebiete.* Springer-Verlag, 1987.

[Bi06] O. Biquard. *Asymptotically Symmetric Einstein Metrics*, volume 13 of *SMF/AMS Texts and Monographs*, 2006.

[Br93] T. P. Branson. *The functional determinant*, volume 4 of *Lecture Notes Series.* Seoul National University Research Institute of Mathematics Global Analysis Research Center, 1993.

[Br95] T. P. Branson. Sharp inequalities, the functional determinant, and the complementary series. *Trans. Amer. Math. Soc.* **347** no. 10 (1995), 3671–3742.

[Br96] T. P. Branson. An anomaly associated with 4-dimensional quantum gravity. *Comm. Math. Phys.*, **178** no.2 (1996), 301–309.

[Br05] T. P. Branson. *Q*-curvature and spectral invariants. *Rend. Circ. Mat. Palermo (2) Suppl.* **75** (2005), 11–55.

[BCY92] T. P. Branson, Sun-Yung A. Chang and P. Yang. Estimates and extremals for zeta function determinants on four-manifolds. *Comm. Math. Phys.* **149** no.2 (1992), 241–262.

[BGP95] T. P. Branson, P. Gilkey and J. Pohjanpelto. Invariants of locally conformally flat manifolds. *Trans. Amer. Math. Soc.* **347** no. 3 (1995), 939–953.

[BG08] T. P. Branson and A. R. Gover. Origins, applications and generalisations of the Q-curvature. *Acta Appl. Math.* **102** no. 2-3 (2008), 131–146.

[BO86] T. P. Branson and B. Ørsted. Conformal indices of Riemannian manifolds. *Compositio Math.* **60** no.3 (1986), 261–293.

[BO88] T. P. Branson and B. Ørsted. Conformal deformation and the heat operator. *Indiana Univ. Math. J.* **37** no. 1 (1988), 83–110.

[BO91a] T. P. Branson and B. Ørsted. Conformal geometry and global invariants. *Diff. Geom. Appl.* **1** (1991), 279–308.

[BO91b] T. P. Branson and B. Ørsted. Explicit functional determinants in four dimensions. *Proc. Amer. Math. Soc.* **113** no. 3 (1991), 669–682.

[B03] S. Brendle, Global existence and convergence for a higher order flow in conformal geometry. *Ann. of Math. (2)* **158** no. 1 (2003), 323–343.

[Bry06] R. Bryant. Conformal Geometry and 3-plane fields on 6-manifolds. In *Developments of Cartan Geometry and Related Mahematical Problems*, volume 1502 of *RIMS Symposium Proceedings*, 1–15, 2006.

[BDS77] D. Burns, K. Diederich and S. Snider. Distinguished curves in pseudoconvex boundaries. *Duke Math. J.* **44** (1977), 407–431.

[CD01] D. Calderbank and T. Diemer. Differential invariants and curved Bernstein-Gelfand-Gelfand sequences. *J. Reine und Angew. Math.* **537** (2001), 67–103. `http://arxiv.org/abs/math/0001158v3`

[Ca02] A. Čap. Parabolic geometry, CR-tractors and the Fefferman construction. *Diff. Geom. Appl.* **17** (2002), 123–138.

[Ca06] A. Čap. Two constructions with parabolic geometries. *Rend. Circ. Mat. Palermo (2) Suppl.*, **79**, 1–37, 2006. ESI-preprint No. 1645, 2005.

[CS09] A. Čap and J. Slovák. *Parabolic Geometries I. Background and General Theory*, volume 154 of *Mathematical Surveys and Monographs*, AMS 2009.

[CSS97] A. Čap, J. Slovák and V.Souček. Invariant operators on manifolds with almost hermitian symmetric structures. II. Normal Cartan connections. *Acta. Math. Univ. Comen. New Ser.* **66** (1997), 203–220.

[CSS01] A. Čap, J. Slovák and V. Souček. Bernstein-Gelfand-Gelfand sequences. *Annals of Math.* **154** (2001), 97–113.

[CY02] Sun-Yung A. Chang and P. Yang. Non-linear partial differential equations in conformal geometry. In *Proceedings of the International Congress of Mathematicians, Vol. I (Beijing, 2002)*, pp. 189–207, 2002.

[CGY03] Sun-Yung A. Chang, M. Gursky and P. Yang. A conformally invariant sphere theorem in four dimensions. *Publ. Math IHES* **98** (2003), 105-143.

[CQY08] Sun-Yung A. Chang, P. Yang and J. Qing, On the renormalized volumes for conformally compact Einstein manifolds. *J. Math. Sciences* **149** no. 6 (2008), 1755–1769. `http://arxiv.org/abs/math/0512376v2`

[C04] Sun-Yung A. Chang. *Non-linear elliptic equations in conformal geometry.* Zürich Lectures in Advanced Mathematics. European Mathematical Society (EMS), Zürich, 2004.

[C05] Sun-Yung A. Chang. Conformal invariants and partial differential equations. *Bull. Amer. Math. Soc. (N.S.)* **42** no. 3 (2005), 365–393 (electronic).

[CEOY08] Sun-Yung A. Chang, M. Eastwood, B. Ørsted and P. Yang. What is Q-curvature? *Acta Appl. Math.* **102** (2008), 119–125.

[CF08] Sun-Yung A. Chang and Hao Fang. A class of variational functionals in conformal geometry. *Intern. Math. Research Notices* **008** (2008). `http://arxiv.org/abs/0803.0333v1`

[CY77] S. Y. Cheng and S. T. Yau. On the regularity of the Monge-Ampère equation $\det(\partial^2 u/\partial x_i \partial x_j) = F(x, u)$, *Comm. Pure Appl. Math.* **30** no.1 (1977), 41–68.

[D99] P. Deligne, P. Etingof, D. S. Freed, L. C. Jeffrey, D. Kazhdan, J. W. Morgan, D. R. Morrison and E. Witten, editors. *Quantum fields and strings: a course for mathematicians. Vol. 1, 2.* AMS, Providence, RI, 1999.

[DS93] S. Deser and A. Schwimmer. Geometric classification of conformal anomalies in arbitrary dimensions. *Phys. Lett. B* **309** no. 3-4 (1993), 279–284. `http://arxiv.org/abs/hep-th/9302047v1`

[HF04] E. D'Hoker and D. Freedman. Supersymmetric gauge theories and the ADS/CFT correspondence. In *Strings, branes and extra dimensions. TASI 2001*, pp. 3–158. World Sci. Publ., NJ, 2004. `http://arxiv.org/abs/hep-th/0201253v2`

[DM08] Z. Djadli and A. Malchiodi. Existence of conformal metrics with constant Q-curvature. *Ann. of Math. (2)*, **168** no. 3 (2008), 813–858. `http://arxiv.org/abs/math/0410141v3`

[DO01] A. J. Di Scala and C. Olmos. The geometry of homogeneous submanifolds of hyperbolic space. *Math. Zeitschrift* **237** (2001), 199–209.

[DL08] A. J. Di Scala and T. Leistner. Connected subgroups of $SO(2, n)$ acting irreducibly on $\mathbb{R}^{2,n}$. To appear in *Israel J. of Math.*. `http://arxiv.org/abs/0806.2586v1`

[DT06] S. Dragomir and G. Tomassini. *Differential Geometry and Analysis on CR manifolds*, volume 246 of *Progress in Mathematics*, Birkhäuser Verlag 2006.

[DNP86] M. J. Duff, B. Nilsson and C. N. Pope. Kaluza-Klein supergravity. *Phys. Rep.* **130** (1986), 1-142.

[ES85] M. Eastwood and M. Singer. A conformally invariant Maxwell gauge. *Physics Letters* **107A** no. 2 (1985), 73–74.

[ES03] M. Eastwood and J. Slovák. A primer on Q-curvature. The American Institute of Mathematics, 2003. `http://www.aimath.org/WWN/confstruct/confstruct.pdf`.

[E05] M. Eastwood. Higher symmetries of the Laplacian. *Annals of Math.* **161** (2005), 1645–1665. `http://arxiv.org/abs/hep-th/0206233v1`

[F07] C. Falk. *Konforme Berandung vollständiger Anti-deSitter-Mannigfaltigkeiten.* Diplom-Arbeit, Humboldt-Universität, Berlin, 2007.

[FJ] C. Falk and A. Juhl, Universal recursive formulae for Q-curvature. To appear in *Crelle's Journal.* http://arxiv.org/abs/0804.2745v2

[Fa86] F. Farris. An intrinsic construction of Fefferman's CR metric. *Pacific J. Math.* **123** (1986), 33–45.

[F76] C. Fefferman. Monge-Ampere equations, the Bergman kernel, and geometry of pseudoconvex domains. *Annals of Math.* **103** (1976), 395–416; **104** (1976), 393–394.

[FG85] C. Fefferman and C. R. Graham. Conformal Invariants. In 'Elie Cartan et les Mathematiques d'Adjourd'hui.' *Astérisque* Numero Hors Serie (1985), 95–116.

[FG02] C. Fefferman and C. R. Graham. Q-curvature and Poincaré metrics. *Math. Res. Lett.* **9** no. 2-3 (2002), 139–151. http://arxiv.org/abs/math/0110271v1

[FG07] C. Fefferman and C. R. Graham. The ambient metric. http://arxiv.org/abs/0710.0919v2

[Fe05] L. Fehlinger. *Holonomie konformer Cartan-Zusammenhänge.* Diplom-Arbeit, Humboldt-Universität, Berlin, 2005.

[Fr81] T. Friedrich. Der erste Eigenwert des Dirac-Operators einer kompakten Riemannschen Mannigfaltigkeit nichtnegativer Skalarkrümmung. *Math. Nachrichten* **97** (1980), 117–146.

[Fr00] T. Friedrich. *Dirac Operators in Riemannian Geometry.* Graduate Studies in Mathematics **25**, AMS, 2000.

[GL08] A. Galaev and T. Leistner. Holonomy groups of Lorentzian manifolds: classification, examples, and applications. In: D. Alekseevsky, H. Baum (eds) *Recent Developments in pseudo-Riemannian Geometry*, 53–96. EMS Publishing House (2008).

[Gi09] N. Ginoux. *The Dirac Spectrum*, volume 1976 of *Lecture Notes in Mathematics.* Springer-Verlag, 2009.

[G06] A. R. Gover. Laplacian operators and Q-curvature on conformally Einstein manifolds. *Math. Ann.* **336** no. 2 (2006), 311–334. http://arxiv.org/abs/math/0506037v3

[GH04] A. R. Gover and K. Hirachi. Conformally invariant powers of the Laplacian – a complete nonexistence theorem. *J. Amer. Math. Soc.* **17** no. 2 (2004), 389–405. http://arxiv.org/abs/math/0304082v2

[GP03] A. R. Gover and L. Peterson. Conformally invariant powers of the Laplacian, Q-curvature, and tractor calculus. *Comm. Math. Phys.* **235** no. 2 (2003), 339–378. http://arxiv.org/abs/math-ph/0201030v3

[Gr87] C.R. Graham. On Sparling's characterization of Fefferman metrics. *Amer. J. Math.* **109** (1987), 853–874.

[G92] C. R. Graham. Conformally invariant powers of the Laplacian. II. Nonexistence. *J. London Math. Soc. (2)* **46** no. 3 (1992), 566–576.

[G00] C. R. Graham. Volume and area renormalizations for conformally compact Einstein metrics. In *The Proceedings of the 19th Winter School "Geometry and Physics" (Srní, 1999)*. Rend. Circ. Mat. Palermo (2) Suppl. No. **63** (2000), 31–42. `http://arxiv.org/abs/math/9909042v1`

[G09] C. R. Graham. Extended obstruction tensors and renormalized volume coefficients. *Advances in Math.* **220** no. 6 (2009), 1956–1985. `http://arxiv.org/abs/0810.4203v1`

[GH05] C. R. Graham and K. Hirachi. The ambient obstruction tensor and Q-curvature. In *AdS/CFT correspondence: Einstein metrics and their conformal boundaries*, volume 8 of *IRMA Lect. Math. Theor. Phys.*, pp. 59–71. Eur. Math. Soc., Zürich, 2005. `http://arxiv.org/abs/math/0405068v1`

[GJMS92] C. R. Graham, R. Jenne, L. J. Mason and G. A. J. Sparling. Conformally invariant powers of the Laplacian. I. Existence. *J. London Math. Soc. (2)* **46** no. 3 (1992), 557–565.

[GJ07] C. R. Graham and A. Juhl. Holographic formula for Q-curvature. *Advances in Math.* **216** no. 2 (2007), 841–853. `http://arxiv.org/abs/0704.1673v1`

[GZ03] C. R. Graham and M. Zworski. Scattering matrix in conformal geometry. *Invent. Math.* **152** no. 1 (2003), 89–118. `http://arxiv.org/abs/math/0109089v1`

[G04] A. Gray. *Tubes*, volume 221 of *Progress in Mathematics.* Birkhäuser Verlag, second edition, 2004.

[Gu09] M. Gursky. PDEs in conformal geometry. in *Lecture Notes in Mathematics* **1977** (2009), pp. 1–33.

[Hab92] K. Habermann. *Twistor-Spinoren auf Riemannschen Mannigfaltigkeiten und deren Nullstellen.* PhD-Thesis, Humboldt-Universität, Berlin, 1992.

[Ha09] M. Hammerl. *Natural Prolongations of BGG-Operators.* PhD thesis, Universität Wien, 2009.

[HaS09] M. Hammerl and K. Sagerschnig. Conformal structures associated to generic rank 2 distibutions on 5-manifolds - Characterization and Killing field decomposition. SIGMA, *Special Issue Elie Cartan and Differential Geometry*, 2009. `http://arxiv.org/abs/0908.0483v1`

[H78] S. Helgason. *Differential Geometry, Lie Groups, and Symmetric Spaces.* Academics Press, 1978.

[H08] S. Helgason. *Geometric analysis on symmetric spaces*, volume 39 of *Mathematical Surveys and Monographs.* AMS, Providence, RI, second edition, 2008.

[HS98] M. Henningson and K. Skenderis. The holographic Weyl anomaly. *JHEP* **7** paper 23 (electronic), 1998. `http://arxiv.org/abs/hep-th/9806087v2`

[J09a] A. Juhl. On conformally covariant powers of the Laplacian. Submitted. `http://arxiv.org/abs/0905.3992v2`

[J09b] A. Juhl. *Families of Conformally Covariant Differential Operators, Q-Curvature and Holography*, volume 275 of *Progress in Mathematics.* Birkhäuser Verlag, 2009.

[J09c] A. Juhl and C. Krattenthaler, Summation formulas for GJMS-operators and Q-curvatures on the Möbius sphere. `http://arxiv.org/abs/0910.4840v1`

[J09d] A. Juhl, On Branson's Q-curvature of order eight. (In preparation.)

[Ka99] I. Kath. *Killing Spinors on Pseudo-Riemannian Manifolds.* Habilitationsschrift, Humboldt-Universität Berlin, 1999.

[K72] S. Kobayashi. *Transformation groups in Differential Geometry.* Springer 1972.

[KN63] S. Kobayashi and K. Nomizu. *Foundations of Differential Geometry I.* J. Wiley and Sons Inc. 1963.

[KS06] M. Kontsevich and Y. Suhov, On Malliavin measures, SLE and CFT. `http://arxiv.org/abs/math-ph/0609056v1`

[K88] W. Kühnel. Conformal Transformations between Einstein Spaces. In: *Conformal Geometry*, R. S. Kulkarni, U. Pinkall (Eds.), Vieweg-Verlag 1988, 105–146.

[KR94] W. Kühnel and H.-B. Rademacher. Twistor spinors with zeros. *Int. J. Math.* **5** (1994), 877–895.

[KR96] W. Kühnel and H.-B. Rademacher. Twistor spinors and gravitational instantons. *Lett. Math. Phys.* **38** (1996), 411–419.

[KR97] W. Kühnel and H.-B. Rademacher. Conformal completion of $U(n)$-invariant Ricci-flat Kähler metrics at infinity. *Z. Anal. Anwend.* **16** no.1 (1997), 113–117.

[KR98] W. Kühnel and H.-B. Rademacher. Asymptotically Euclidean manifolds and twistor spinors. *Comm. Math. Phys.* **196** (1998), 67–76.

[KR00] W. Kühnel and H.-B. Rademacher. Asymptotically Euclidean ends of Ricci flat manifolds, and conformal inversions. *Math. Nachr.* **219** (2000), 125–134.

[Ku97] M. Kuranishi. CR-structures and Fefferman's conformal structures. *Forum Math.* **9** (1997), 127–164.

[ML89] H. B. Lawson and M.-L. Michelsohn. *Spin Geometry.* Princeton University Press, 1989.

[Lee86] J. M. Lee. The Fefferman metric and pseudohermitian invariants. *Trans. AMS* **296** (1986), 411–429.

[LP87] J. M. Lee and T. H. Parker. The Yamabe problem. *Bull. Amer. Math. Soc. (N.S.)* **17** no. 1 (1987), 37–91.

[Lei03] T. Leistner. *Holonomy and parallel spinors in Lorentzian geometry.* PhD-Thesis, Humboldt-Universität, Berlin, 2003.

[Lei06] T. Leistner. Conformal holonomy of C-spaces, Ricci-flat, and Lorentzian manifolds. *Diff. Geom. Appl.* **24** (2006), 458–478. `http://arxiv.org/abs/math/0501239v3`

[Lei07] T. Leistner. On the classification of Lorentzian holonomy groups. *Journal of Differential Geometry* **76** no. 3 (2007), 423–484.

[LN09] T. Leistner and P. Nurowski. Conformal structures with $G_{2(2)}$-ambient metrics. `http://arxiv.org/abs/0904.0186v2`.

[L01] F. Leitner. *The Twistor Equation in Lorentzian Spin Geometry.* PhD-Thesis, Humboldt-Universität, Berlin, 2001.

[L05a] F. Leitner. Conformal Killing forms with normalization condition. *Rend. Circ. Mat. Palermo*, suppl. Ser II, **75** (2005), 279–292.

[L05b] F. Leitner. About complex structures in conformal tractor calculus. `http://arxiv.org/abs/math/0510637v2`

[L07] F. Leitner. *Applications of Cartan and Tractor Calculus to Conformal and CR-Geometry.* Habilitationsschrift, Stuttgart, 2007.

[L08] F. Leitner. A remark on unitary conformal holonomy. In: *Symmetries and overdetermined systems of partial differential equations*, vol. 144 of *IMA Vol. Math. Appl.*, 445–460, Springer, 2008

[Lew91] J. Lewandowski. Twistor equation on a curved spacetime. *Class. Quant. Grav.* **8** (1991), 11–17.

[Li88] A. Lichnerowicz. Killing spinors, twistor-spinors and Hijazi inequality. *J Geom. Phys.* **5** (1988), 2–18.

[M07] A. Malchiodi. Conformal metrics with constant Q-curvature. *SIGMA Symmetry Integrability Geom. Methods Appl.* **3**, Paper 120, 11, 2007. `http://arxiv.org/abs/0712.2123v1`.

[M98] J. Maldacena. The large N limit of superconformal field theories and supergravity. *Adv. Theor. Math. Phys.* **2** no. 2 (1998) 231–252. `http://arxiv.org/abs/hep-th/9711200v3`

[MM87] R. Mazzeo and R. Melrose, Meromorphic extension of the resolvent on complete spaces with asymptotically constant negative curvature. *J. Funct. Anal.* **75** (1987) 260-310.

[Me95] R. Melrose. *Geometric scattering theory.* Stanford Lectures. Cambridge University Press, 1995.

[Nd07] C. B. Ndiaye. Constant Q-curvature metrics in arbitrary dimension. *J. Funct. Anal.* **251** no. 1 (2007), 1–58.

[Ni89] P. Nicholls. *The Ergodic Theory of Discrete Groups*, volume 143 of *London Mathematical Society Lecture Note Series.* Cambridge University Press, 1989.

[Nu05] P. Nurowski. Differential equations and conformal structures. *Journ. Geom. Phys.* **55** no.1 (2005), 19–49. `http://arxiv.org/abs/math/0406400v3`

[Nu08] P. Nurowski. Conformal structures with explicit ambient metrics and conformal G_2 holonomy. In *Symmetries and overdetermined systems of partial differential equations*, volume 144 of *IMA Vol. Math. Appl.*, 515–526, Springer, New York, 2008. `http://arxiv.org/abs/math/0701891v2`

[OV07] J. F. O'Farrill and B. Vercnocke. Killing superalgebra deformations of ten-dimensional supergravity backgrounds. Bert *Class. Quant. Grav.* **24** (2007), 6041–6070. `http://arxiv.org/abs/0708.3738v1`

[OHMS09] J. F. O'Farrill, E. Hackett-Jones, G. Moutsopoulos and J. Simón. On the maximal superalgebras of supersymmetric backgrounds. *Class. Quant. Grav.* **26** no. 3 (2009). http://arxiv.org/abs/0809.5034v1

[O82] E. Onofri. On the positivity of the effective action in a theory of random surfaces. *Comm. Math. Physics* **86** (1982), 321–326.

[OPS88a] B. Osgood, R. Phillips and P. Sarnak. Extremals of determinants of Laplacians. *J. Funct. Anal.* **80** no. 1 (1988), 148–211.

[OPS88b] B. Osgood, R. Phillips and P. Sarnak. Compact isospectral sets of surfaces. *J. Funct. Anal.* **80** no. 1 (1988), 212–234.

[Pa83] S. Paneitz. A quartic conformally covariant differential operator for arbitrary pseudo-Riemannian manifolds (summary). *SIGMA Symmetry Integrability Geom. Methods Appl.* **4** Paper 036, 3, 2008. http://arxiv.org/abs/0803.4331v1

[PR86] R. Penrose and W. Rindler: *Spinors and Space-time II.* Cambridge Universiy Press, 1986.

[Po81] A. M. Polyakov. Quantum geometry of bosonic strings. *Phys. Lett. B* **103** no. 3 (1981), 207–210.

[RS71] D. B. Ray and I. M. Singer. R-torsion and the Laplacian on Riemannian manifolds. *Advances in Math.* **7** (1971), 145–210.

[R84] R. Riegert. A nonlocal action for the trace anomaly. *Phys. Lett. B* **134** no. 1-2 (1984), 56–60.

[Sch85] W. Schmid. Boundary value problems for group invariant differential equations. *Astérisque* Numero Hors Serie (1985), 311–321.

[Sch97] M. Schottenloher. *A Mathematical Introduction to Conformal Field Theory.* Springer-Verlag, 1997.

[Sh97] R. W. Sharpe. *Differential geometry, Cartan's generalization of Klein's Erlangen program* volume 166 of *Graduate Texts in Mathematics.* Springer-Verlag, 1997.

[Sk02] K. Skenderis. Lecture notes on holographic renormalization. *Classical Quantum Gravity* **19** no. 22 (2002) 5849–5876. http://arxiv.org/abs/hep-th/0209067v2

[Sp85] G. Sparling. Twistor theory and the characterization of Fefferrman's conformal structures (preprint). Univ. Pittsburg, 1985.

[V00] J. Viaclovsky. Conformal geometry, contact geometry, and the calculus of variations. *Duke Math. J.* **101** no. 2 (2000), 283–316.

[V06] J. Viaclovsky. Conformal geometry and fully nonlinear equations. http://arxiv.org/abs/math/0609158v1

[Wan89] McKenzie Y. Wang. Parallel spinors and parallel forms. *Ann. Glob. Anal. and Geom.* **7** no.1 (1989), 59–68.

[Wi98] E. Witten. Anti de Sitter space and holography. *Adv. Theor. Math. Phys.* **2** no. 2 (1998), 253–291. http://arxiv.org/abs/hep-th/9802150v2

Index

Oberwolfach Seminars (OWS)

BIRKHÄUSER

The workshops organized by the *Mathematisches Forschungsinstitut Oberwolfach* are intended to introduce students and young mathematicians to current fields of research. By means of these well-organized seminars, also scientists from other fields will be introduced to new mathematical ideas. The publication of these workshops in the series *Oberwolfach Seminars* (formerly *DMV seminar*) makes the material available to an even larger audience.

OWS 40: Baum, H. / Juhl, A., Conformal Differential Geometry. Q-Curvature and Conformal Holonomy (2010). ISBN 978-3-7643-9908-5

OWS 39: Drton, M. / Sturmfels, B. / Sullivant, S., Lectures on Algebraic Statistics (2008). ISBN 978-3-7643-8904-8

How does an algebraic geometer studying secant varieties further the understanding of hypothesis tests in statistics? Why would a statistician working on factor analysis raise open problems about determinantal varieties? Connections of this type are at the heart of the new field of "algebraic statistics". In this field, mathematicians and statisticians come together to solve statistical inference problems using concepts from algebraic geometry as well as related computational and combinatorial techniques. The goal of these lectures is to introduce newcomers from the different camps to algebraic statistics. The introduction will be centered around the following three observations: many important statistical models correspond to algebraic or semi-algebraic sets of parameters; the geometry of these parameter spaces determines the behaviour of widely used statistical inference procedures; computational algebraic geometry can be used to study parameter spaces and other features of statistical models.

OWS 38: Bobenko, A.I. / Schröder, P. / Sullivan, J.M. / Ziegler, G.M. (Eds.), Discrete Differential Geometry (2008). ISBN 978-3-7643-8620-7

Discrete differential geometry is an active mathematical terrain where differential geometry and discrete geometry meet and interact. It provides discrete equivalents of the geometric notions and methods of differential geometry, such as notions of curvature and integrability for polyhedral surfaces. Current progress in this field is to a large extent stimulated by its relevance for computer graphics and mathematical physics. This collection of essays, which documents the main lectures of the 2004 Oberwolfach Seminar on the topic, as well as a number of additional contributions by key participants, gives a lively, multi-facetted introduction to this emerging field.

OWS 37: Galdi, G.P. / Rannacher, R. / Robertson, A.M. / Turek, S., Hemodynamical Flows (2008). ISBN 978-3-7643-7805-9

This book surveys results on the physical and mathematical modeling as well as the numerical simulation of hemodynamical flows, i.e., of fluid and structural mechanical processes occurring in the human blood circuit. The topics treated are continuum mechanical description, choice of suitable liquid and wall models, mathematical analysis of coupled models, numerical methods for flow simulation, parameter identification and model calibration, fluid-solid interaction, mathematical analysis of piping systems, particle transport in channels and pipes, artificial boundary conditions, and many more.

OWS 36: Cuntz, J. / Meyer, R. / Rosenberg, J.M., Topological and Bivariant K-theory (2007). ISBN 978-3-7643-8398-5

Topological K-theory is one of the most important invariants for noncommutative algebras. Bott periodicity, homotopy invariance, and various long exact sequences distinguish it from algebraic K-theory. We describe a bivariant K-theory for bornological algebras, which provides a vast generalization of topological K-theory.

OWS 35: Itenberg, I. / Mikhalkin, G. / Shustin, E., Tropical Algebraic Geometry (2007). ISBN 978-3-7643-8309-1

OWS 34: Lieb, E.H. / Seiringer, R. / Solovej, J.P. / Yngvason, J., The Mathematics of the Bose Gas and its Condensation (2005). ISBN 978-3-7643-7336-8

OWS 33: Kreck, M. / Lück, W., The Novikov Conjecture: Geometry and Algebra (2004). ISBN 978-3-7643-7141-8